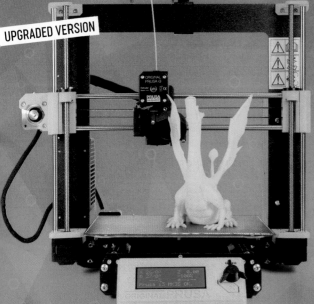
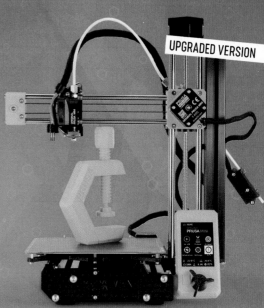

CONTENTS

16

28

Zach Cordner, Nathaniel Taylor, Michael Krzyzaniak, Eben Kouao, Katie Rosa Marchese, Lenny Leiter, Johan von Konow, Billie Ruben

Make:®

"You have to remember that once upon a time a violin was technology, once upon a time an organ was technology. Those things were all built and created by people who were working at the cutting edge of the technologies of their time." –Brian Eno

PRESIDENT
Dale Dougherty
dale@make.co

VP, PARTNERSHIPS
Todd Sotkiewicz
todd@make.co

EDITORIAL

EXECUTIVE EDITOR
Mike Senese
mike@make.co

SENIOR EDITORS
Keith Hammond
keith@make.co

Caleb Kraft
caleb@make.co

PRODUCTION MANAGER
Craig Couden

CONTRIBUTING EDITOR
William Gurstelle

CONTRIBUTING WRITERS
Simon Begg, Elizabeth Clark, Larry Cotton, Kelly Egan, Greg Gilman, Johan von Konow, Eben Kouao, Michael Joseph Krzyzaniak, Tod Kurt, Helen Leigh, Mario Marchese, Carsten Mayer, Peter Mišenko, Patrick Peters, Benjamin Prescher, Billie Ruben, Dan Schneiderman, Jeff Shaw, Andreas Spiess, Nathaniel Taylor, Tom Whitwell

CONTRIBUTING ARTIST
Zach Cordner

MAKE.CO

ENGINEERING MANAGER
Alicia Williams

WEB APPLICATION DEVELOPER
Rio Roth-Barreiro

BOOKS

BOOKS EDITOR
Patrick DiJusto

DESIGN

CREATIVE DIRECTOR
Juliann Brown

GLOBAL MAKER FAIRE

MANAGING DIRECTOR, GLOBAL MAKER FAIRE
Katie D. Kunde

MAKER RELATIONS
Siana Alcorn

GLOBAL LICENSING
Jennifer Blakeslee

MARKETING

DIRECTOR OF MARKETING
Gillian Mutti

LEARNING LABS

DIRECTOR OF LEARNING
Nancy Otero

OPERATIONS

ADMINISTRATIVE MANAGER
Cathy Shanahan

ACCOUNTING MANAGER
Kelly Marshall

OPERATIONS MANAGER & MAKER SHED
Rob Bullington

PUBLISHED BY

MAKE COMMUNITY, LLC
Dale Dougherty

Copyright © 2021
Make Community, LLC. All rights reserved.
Reproduction without permission is prohibited.
Printed in the USA by Schumann Printers, Inc.

Comments may be sent to:
editor@makezine.com

Visit us online:
make.co

Follow us:
🐦 @make @makerfaire @makershed
📘 makemagazine
📷 makemagazine
▶ makemagazine
🎮 twitch.tv/make
📌 makemagazine

Manage your account online, including change of address:
makezine.com/account
866-289-8847 toll-free in U.S. and Canada
818-487-2037,
5 a.m.–5 p.m., PST
cs@readerservices.makezine.com

Make: Community

Support for the publication of *Make:* magazine is made possible in part by the members of Make: Community. Join us at make.co.

CONTRIBUTORS
What was your most epic musical performance?

Eben Kouao
London, United Kingdom
(Live Streaming With Raspberry Pi)
This has definitely got to be an evening Jazz Live Band. No electronics, simply a saxophone and a drumset.

Elizabeth Clark
Boston, MA
(Glocken' Roll)
Every show I played as a teenager felt epic and my band always felt on top of the world afterwards.

Helen Leigh
Portland, OR
(Aural Innovations)
At Musikmesse in Frankfurt last year I performed as part of an orchestra of music hackers and artists playing DIY instruments, including cello sounds that were bounced off the moon, an AI choir and my own circuit sculpture harp.

Issue No. 76, Spring 2021. *Make:* (ISSN 1556-2336) is published quarterly by Make Community, LLC, in the months of February, May, Aug, and Nov. Make Community is located at 150 Todd Road, Suite 200, Santa Rosa, CA 95407. SUBSCRIPTIONS: Send all subscription requests to *Make:*, P.O. Box 17046, North Hollywood, CA 91615-9588 or subscribe online at makezine.com/offer or via phone at (866) 289-8847 (U.S. and Canada); all other countries call (818) 487-2037. Subscriptions are available for $34.99 for 1 year (4 issues) in the United States; in Canada: $43.99 USD; all other countries: $49.99 USD. Periodicals Postage Paid at San Francisco, CA, and at additional mailing offices. POSTMASTER: Send address changes to *Make:*, P.O. Box 17046, North Hollywood, CA 91615-9588. Canada Post Publications Mail Agreement Number 41129568. CANADA POSTMASTER: Send address changes to: Make Community, PO Box 456, Niagara Falls, ON L2E 6V2

Make: Projects

GROWING BEYOND EARTH
—— MAKER CHALLENGE ——
YEAR 2: AUTONOMOUS PLANT SYSTEMS

FAIRCHILD TROPICAL BOTANIC GARDEN · MOONLIGHTER FABLAB · NATION OF MAKERS · adafruit

Year 2 Challenge: Autonomous Growing Systems
Maintaining Plants Without Human Intervention

Fairchild Tropical Botanic Garden is working with NASA to help improve food production for long-duration space travel. Together, we're calling on makers across America to submit new designs for gardening systems to be used aboard spacecraft. **Year 1** focused on the volumentric design of the growth chamber (see the winners below). This year's challenge is to develop a smart system that might include sensors, cameras, or automated controllers for lighting, watering, and air circulation within a 50 cubic centimeter growing environment. You'll plant predetermined seeds and then let them grow without human interaction over a 30 day period. Throughout the challenge you'll have chances to consult experts at the botanical garden and NASA to help inform your design decisions. Your ideas may go on to improve the plant chambers on the International Space Station — and beyond!

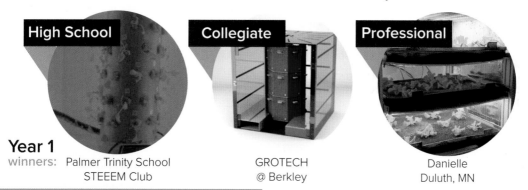

High School

Collegiate

Professional

Year 1 winners:
Palmer Trinity School STEEEM Club

GROTECH @ Berkley

Danielle Duluth, MN

Save The Dates
Dec 1 – First webinar
Jan 12 – Chat with NASA
Feb 9 – Progress/critique
Mar 9 – Chat with experts
Maker Faire Miami – Present Phase 1
May 11 – Chat with experts
NOMCON – Finalist presentations

Go to **makeprojects.com/growing-beyond-earth** **to learn more and register for the challenge!**

LEARN MORE!

This project is based upon work supported by NASA under award No. 80NSSC18K1225. Any opinions, findings and conclusions or recommendations expressed in this material are those of the authors and do not necessarily reflect the views of the National Aeronautics and Space Administration.

Calling Policy Makers

by Dale Dougherty, *Make:* President

Years ago, I gave a talk in Berlin at a BMW-sponsored outdoor event and I said that the maker movement was *apolitical*. After my talk, a young person approached me and questioned that statement. "Everything's political," he said, then shrugged his shoulders and walked away. These days, everything indeed seems to be political.

What I meant by apolitical was that making is something that anyone could support, regardless of political persuasion, which is how we talk about education, the arts, as well as agriculture and manufacturing. It can be something that brings us together, not divides us. However, there is also a difference between politics and policy. If politics is about elections, then policy speaks to what the government should or should not do. What might the maker community ask the government to do?

Sabrina Merlo, who worked on Maker Faire for many years, has been working with Open Source Medical Supplies on local response efforts for Covid-19. She believes that the maker response could have been even more effective if the government had played a stronger, coordinating role and provided funding. With a new administration in 2021, Sabrina thought that the maker community needed its own process to develop policy proposals. She reached out to Dorothy Jones-Davis, Executive Director of Nation of Makers, to organize several Maker Town Halls to encourage more people to participate and seek consensus around the best ideas.

Sabrina also learned from working with the Bay Area Bicycle Coalition in a previous job. The bicycle advocacy community had a process for prioritizing their "asks" into a policy platform. They would develop specific platforms for each level of government: federal, state, regional, and local levels. "When members of the Bicycle Coalition visited Capitol Hill, for example, they were prepared with talking points around federal and state policy platforms to discuss with their representatives," she says.

Two of the policy positions that Sabrina has worked on address the role of makerspaces and makers in future emergencies. One is to create a "certification structure for makerspaces so that they are more understood by/available to state emergency response agencies, and also more eligible for emergency response funding." Another is to create "an open source design library" for tools and devices that can be manufactured in makerspaces called a "U.S. Digital Stockpile." Stephanie Santoso and Megan Brewster, both of whom worked in the Obama Administration's Office of Science and Technology Policy, propose that the government might "leverage makerspaces and Fab Labs as local sites for preparing the current and future workforce through real-world maker-centered learning and training." Another is a proposal developed by Dorothy to place Makers-in-Residence in federal agencies. A list of policy proposals can be found at nationofmakers.us/advocacy-policy.

Makers do not need policy or government action to act. That is made clear by the incredible response to Covid-19 by makers in all parts of the world. The pandemic also showed that makerspaces can be organized around a purpose and the combined efforts of members and community volunteers can be directed toward shared goals. One can imagine, as Sabrina does, that a national network of makerspaces could provide infrastructure and talent that can be deployed in times of emergency but also help address economic and social inequities in urban and rural areas. There's an opportunity to think about how our goals align with the national interest, and develop policies that invest in our community so that we are always ready to solve new problems in the future. ●

Adobe Stock - Igor Serazetdinov

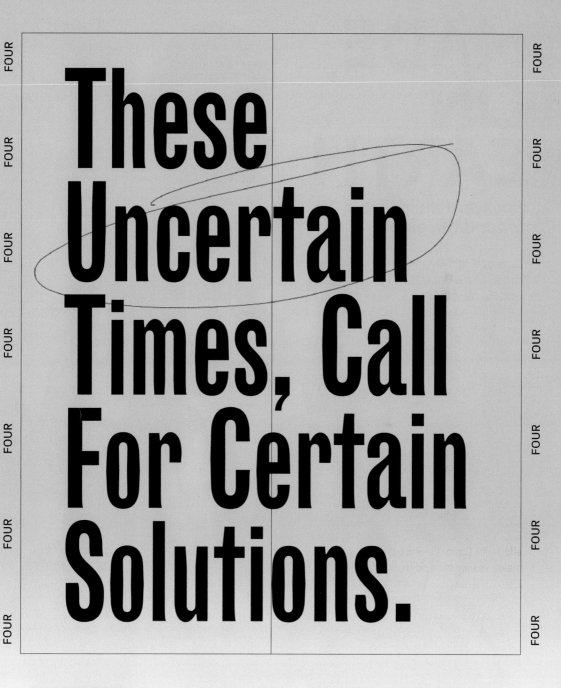

These Uncertain Times, Call For Certain Solutions.

THE FESTIVAL OF URGENT REINVENTIONS

Find out more at <u>thefour.live</u>

THE FESTIVAL OF URGENT REINVENTIONS is a free two-day virtual experience. Part conference, part competition – the festival will be a combination of talks and workshops by prominent change makers, and actionable briefs tackling the world's most urgent and systemic issues.

MADE ON EARTH

Backyard builds from around the globe

Know a project that would be perfect for Made on Earth? Let us know: *editor@makezine.com*

THE BLADE RUNNER

INSTAGRAM.COM/GREGOLIJNYK

It's hard to believe these otherworldly, meticulous robotic sculptures are made out of cardboard, and may be even harder to believe that the Melbourne-based graphic designer spending hundreds of hours crafting each piece is just winging it. "I kind of make them up as I go along. I don't do a sketch," artist **Greg Olijnyk** says. "The whole idea is to get away from the electronic side, to work with the hands and the eye." The majestic creations, occupying Olijnyk's time for the last four years, are the culmination of skills the artist spent 40 years developing, harkening back to a time when graphic design was analogue, requiring a scalpel and raw materials, instead of a computer. A big imagination and a fine eye for detail are essential to assembling these mechanical characters, simply titled by numbers in order of creation, but Olijnyk emphasizes the most important virtue in the process is patience. "It's really having the fortitude to sit there and cut little bricks for hours and days on end."

So far, the 61-year-old artist has made 14 cardboard creations, including mesmerizing miniature apartment buildings for an upcoming series called *The Neighbors*, juxtaposing urban civilization with the wonderfully weird mechanical world he's been developing piece by piece. Olijnyk imagines a gallery exhibition one day, but for now, he's happy utilizing Instagram, and is more interested in the process than any financial payoff. The Vespa bot, named *#7* (opposite page), ranks among his favorites, because cardboard curvature is a more daunting challenge to sculpt, and real-world mechanics require more faithful detail. He wants "people to discover something the longer they look," and *#7* is ripe for staring. At first glance, it's a robot riding a scooter. Look a little harder, and the robot *is* the scooter.

—*Greg Gilman*

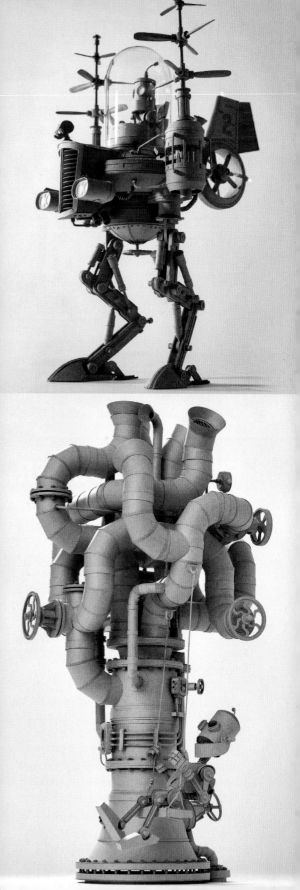

FLIGHT TRAINER MAKEZINE.COM/GO/KIDDY-COPTER

Chuck Helmholdt spent 13 years building his first airplane, a full-size Piper PA-14 replica, with the assistance of his son Tony. He then built three sit-on-top Kiddy Cub plane toys from plans that he modified, adding navigation lights, engine sounds, brakes, throttle, and so on, pulling from his 40 years of piloting experience to add realism to the creations. When Tony, who now works at Tesla, brought his personal electric motorcycles to display at his local Grand Rapids, MI Maker Faire, Chuck joined along, bringing his smaller Piper-inspired, push-behind plane builds to let the younger attendees try out. "I had a great time teaching a little about airplanes and aviation," he says.

That led to his next build. "I had never seen anyone make a kid-sized helicopter before and thought it would be unique and also be a great asset to my Kiddy Cub collection," Helmholdt says. He drew up plans for a bubble-cockpitted design inspired by the classic Bell 47 chopper

seen on *M*A*S*H*, using dimensions that would allow him to bring it into children's museums and hospitals. He and Tony then began hacking away.

Helmholdt completed the Kiddy Copter a year later, and it is as eye-catching as it is interactive. "It has a complete articulating motorized rotor head and the main rotor blades can be tilted when the collective is raised on the flight deck," he says. "There is a complete instrumented flight deck with switches that operate strobes, beacons, engine sounds, and a camera video monitor system. It's designed to be pushed from the tail rotor post where an adult can also use actuators to lift the Kiddy Copter six inches above the ground and simulate a ground effect hover."

Helmholdt estimates the final build tab at $12,000. "Another one would probably cost less because I learned through trial and error what batteries and motors to use, and other fabrication detours." He hopes to bring it back out on parades in the Spring. —*Mike Senese*

FABRICANATOMY

Taxidermy is the art of preserving an animal's dead body for display, but Austria-based artist **Natalia Lubieniecka** doesn't need to kill anything to achieve stunning results. Her game is hunting down the right textiles to depict the cycle of life and death, sculpting colorful pieces inspired by nature from fabric, wire, paint, and other raw materials.

The wide array of creations for sale through her Etsy shop (etsy.com/shop/mysouldesign), which has drawn thousands of visitors since launching in 2014, were initially inspired by a trip to the Natural History Museum in Vienna. She says that its "wonderful collection of taxidermy animals" sparked her line of dissected frogs and rats, insect brooches and pendants, spider sculptures encased in glass, vivid mushrooms, and more recently, human organs. "Hearts and brains are my favorite topic," she says. "The brain and heart are connected together, as are feelings and thoughts. My sculptures show my new rebirth, and perception of the world from a different perspective — a better perspective."

Her work is a reflection of a deep faith in reincarnation and the interconnection of all organisms in nature, which operates on a simple truth — life must die in order to continue. Fungi is a part of that process, as one of the Earth's most effective decomposers, and it's another one of Lubieniecka's favorite subjects to work with. "I love mushrooms," she says, recalling a childhood spent picking them. "I loved the toadstool groups, in particular. They are something from a fairy tale." With over 94,000 Instagram followers, and clients from around the world shelling out hundreds of dollars per piece, it seems she is well on her way to living happily ever after.
                                        —*Greg Gilman*

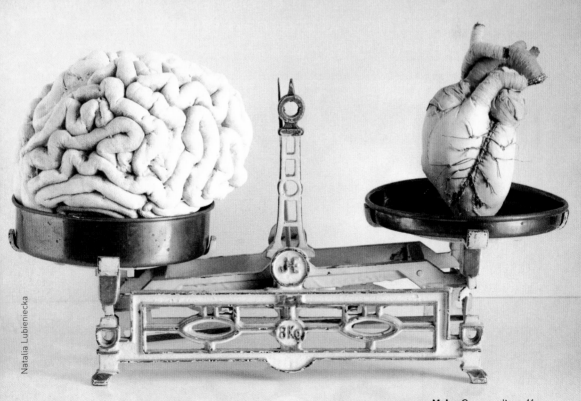

Natalia Lubieniecka

Making a Safer Space

Narwhal Labs was about to open its doors. Then the shutdown hit and the makerspace needed to navigate a new course **Written by Jeff Shaw**

I t was Friday, March 6, 2020 and our new public makerspace in Bristol, Rhode Island, was on target to open in just a few weeks. Nearly three hard months of designing, making, and building furniture for the space, researching, writing and developing policy, building websites, getting people excited on social media, and setting up tools had us anticipating the opening in just a few weeks. The end of the work week came and it was time to go home. We turned off the lights and locked the doors for the weekend. Monday came along much too quickly (the weekends are never long enough!) when, due to the new and rapidly spreading Covid-19 virus, the Governor of Rhode Island declared a state of emergency. Businesses shut down, schools went remote, and workers all over began a massive shift to working from home. We too decided to cancel the upcoming makerspace opening. It wasn't an easy choice, but it was the responsible one.

The spring and summer were bumpy for everyone but Narwhal Labs finally opened to the public in October 2020. For many makerspaces and hackerspaces around the world, unfortunately, the story doesn't have the same happy ending. Some doors have shut indefinitely. Others have closed permanently or struggle to make rent. Many of these wonderful local communities of like-minded tinkerers, hackers, and makers still aren't sure they'll have a physical space to return to, or tools and equipment to use. For those still operating, engagement has declined, and access is limited.

HOLDING THE FORT

For the months following the cancelled opening, the crew at Narwhal Labs focused on content — making entertaining and educational videos to start our new YouTube channel, and making new content for the space's sponsor, TotalBoat (Narwhal Labs is located at their facilities). We made arrangements with some friends to help make videos and social media posts. We even hired a full-time videographer. I see our facility as a test kitchen of sorts — developing techniques and learning, and using digital media as a way to share what we learn. It doesn't matter if you're making a cake, resin

JEFF SHAW (@ideal_grain) is a maker generalist and director of Narwhal Labs, a makerspace in Bristol, Rhode Island.

Adobe - Fiedels, Narwhal Labs

art, or a mobile guitar stage and store — there's still testing to be done, recipes of some kind to be perfected, and lessons to be learned to achieve success with a project. Our crew has been able to work with local friends and those passing through like Xyla Foxlin (IG: @xylafoxlin), Tim Sway (@timsway1), Jessie Jewels (@jessiejewelsart), Paul Jackman (@jackman_works), Troy Conary (@arbortechie), as well as Phillip and Elizabeth Danner (@dannerbuilds and @dannerbuildswifey) on videos. We've also collaborated on live-stream videos with our friends Sami and Cory at AvidCNC, and for the virtual Catskill Mountain Maker's Camp weekend.

Keeping an open mind on how the makerspace gets utilized has given us more reasons to keep pushing forward even when we couldn't open. The tools and equipment at the makerspace were

OPENING SAFELY

crucial for making 3D printed and laser-cut PPE, as so many other makers did when the pandemic broke out and supply chains were struggling. Our backing and expanded multi-function purpose gave us the flexibility to take our time and ensure we made the right decisions when it was appropriate. TotalBoat has used our space for product testing and development, marketing, product photography, and even as a socially distant meeting space. A few months in, we also began to pilot the idea of opening Narwhal Labs by holding "shop nights" — inviting our sponsor's employees to take classes on digifab equipment and learn to use the tools and supplies around the shop.

We came up with some basic safety protocols to allow our makerspace to open and finally give back to the maker community as we intended. Of course, masks that cover the nose and mouth are required for anyone in the building, regular hand washing is encouraged, and hand sanitizer is made available. Members are given their own safety glasses to keep so they aren't being shared. The makerspace is open by appointment only, and is limited to four members in the space at a time. This keeps us under half of the maximum gathering size under the state of Rhode Island's mandates and guidelines as of the writing of this article. Booking appointments also helps us keep track of attendance for contact

The Narwhal Labs Crew Left to Right:
Jeff — Director of Narwhal Labs, Skip — CIO of TotalBoat, Graz — Narwhal Labs Videographer, Kristin — Social Media Coordinator

Nurses and staff at Women & Infants Hospital in Rhode Island wearing face shields and ear savers manufactured at Narwhal Labs

tracing. Not coming in when sick and alerting staff if a member tests positive for Covid-19 for sanitization and contact tracing are common sense rules. We provide an alcohol sanitizing solution in spray bottles, and ask that anyone in the shop sanitize tables, workbenches, touch surfaces on tools, and loaner laptops before and after use. We also sanitize these items when members aren't around. Our work tables are currently set up for only one person per table, and spaced out for a minimum 6 foot distance between any seating position. Membership is discounted during this time to help make up for these challenges. When weather permits, we open the garage door and windows to our workshop to keep air flowing. These changes are working well for us and were reasonable and simple to implement.

We believe that through these process and policy changes we'll be able to remain open through the end of the pandemic and beyond. We're grateful to our local community and partnered brands for their patience and support that have enabled us to take our time to make informed decisions to safely open our makerspace. After the bulk of this article was written, new temporary state orders for recreational facilities have caused us to shut our doors for a couple weeks, but we're confident in our ability to safely operate when we welcome our members back. The spaces I mentioned before — the ones that might not be doing so well right now: they need your help. If you represent a business in the maker community, consider sponsoring and supporting local makerspaces in your local area that need it — content partnerships and equipment can provide a valuable marketing opportunity, potentially a tax write-off, and help keep these local spaces alive. To individual makers: Find your local, community makerspace or hackerspace and ask how you can help. Making a donation is great, but consider becoming a member to contribute and help keep our local communities of makers going for a future generation. ◐

- General information and membership: narwhallabs.com
- Technical information and Covid safety: wiki.narwhallabs.com
- All Narwhal Labs design files for CNC, laser and 3D printing, CAM tool libraries, and more: github.com/narwhallabs
- Narwhal Labs on YouTube: youtube.com/narwhallabs
- Narwhal Labs on Instagram: @narwhallabs

Narwhal Labs, Nursing staff at Women & Infants Hospital

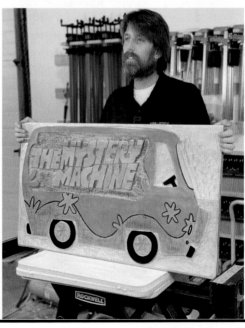

DALE DOUGHERTY
is the president of Make:

Me and My Robot

Odd Jayy wanted a friend so he built a robot — and in doing so, he discovered his community

Written by Dale Dougherty

Jorvon Moss, known on Twitter as @Odd_Jayy, recently showed off his second pair of Magpie goggles with the tweet: "And this is why people called me odd growing up." I reached out to him and he explained that the glasses were "a fun gadget that he put together really quickly over a weekend." He programmed the 3D-printed mechanical device to open and close the irises and raise and lower separate eyebrows, using an Adafruit Trinket and two servo motors. More mechanical than digital, the goggles become part of a costume, and Jorvon becomes his own character. "Every sci-fi movie always had a really cool person with goggles," Moss told me. A video of him wearing the Magpie goggles "became a lot more popular than I thought," he said. His Twitter followers, many of them makers he has gotten to know, cheered him on.

Moss began building robots about five years ago when he was in college. Like a lot of makers, he started doing it before he understood what he was doing, like artists who start drawing, even though they haven't had a lot of training.

"I had no idea about electronics," he says. In his dorm room, the first project he tried was putting electronics into Plushies. "I didn't know about microcontrollers back then. I bought a AA battery pack with 12 AA batteries and a single micro servo. I plugged the power pack directly into the servo and the servo exploded." His roommate freaked out. "He was like, 'what are you doing? You're trying to blow up the room.' I'm like, 'no, I was just trying to do science.'"

Growing up in Compton, California, Moss was raised in a religious household and went to a religious school that didn't teach science. "I'm a D student throughout most of my young life. I just wasn't interested in the work. I was spending most of my time reading something or drawing, and it got me in trouble a lot." He thought: "Hey, I might as well get a good job in something that I love."

He went to the Academy of Art University in San Francisco to become a comic book illustrator. "I wanted to be a penciler," he says. He was learning the traditional arts to draw panels and do basic things like perspective that artists sketch in their sketchbooks.

Moss began working with electronics as a diversion from drawing. "It was very surprising for me just to fall in love with it. After you've been drawing all day in school and then your teachers are putting it in your head that you have to draw every day, all the time, it got to a point where I wanted to do anything else with my free time than draw. Anything else. That's where the hobby of

Zach Cordner

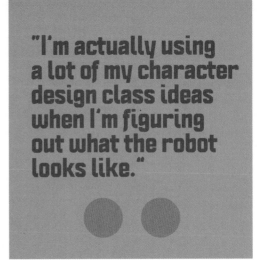

"I'm actually using a lot of my character design class ideas when I'm figuring out what the robot looks like."

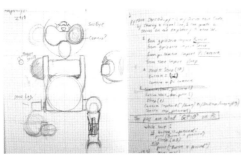

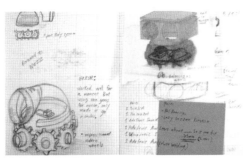

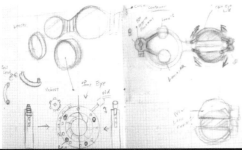

electronics came up. Let me see how I enjoy it. And I ended up really loving it."

With his own projects, there was no pressure on him. "When it came to technology, people are surprised that you can get a simple servo to move," he says. He liked that he would not be judged.

While going to college, he was also working full-time. He recalls that he had a free day when his classes were cancelled. "I was able to go get lunch for the first time," he says, describing a huge lunch room. "I remember I got lunch and went to sit down. All the tables were filled with people." He ended up at a table to eat by himself and it dawned on him: "No one knew who I was. I didn't really have any friends because I was so busy working all the time," he says, looking back on it. "That kind of hurt. It was super lonely."

He headed home with that feeling and watched a Star Wars or a sci-fi movie. He noticed the robots. "How perfect that they have robot best friends in those movies," he said to himself. "That's genius. I will just build myself my own best friend." That is what inspired him to eventually build the robot companion that he'd name Dexter (featured in *Make:* Vol. 73).

After graduating with a degree in illustration, Moss moved back home with his family. "Because I was so focused on graduating, I didn't sit down and work on my portfolio at all." He found various part-time jobs, trying to figure out what he really wanted to do. He kept tinkering.

"I ended up getting a 3D printer. I had to save up for one and it took me a while," he says.

Jorvon Moss, Moheeb Zara

As he started printing things, he posted them on Instagram. "Use the tag #3Dprinting and eventually someone is going to notice you." A person who noticed him said he should join a makerspace. "What's a makerspace?" he asked. He was introduced to a nearby spot in Culver City called Crash Space. Through the makerspace, he found a community. "I had people who were showing me all these new things," he says. "It was cool."

His first robot was called Tuesday, "because that was our meeting day at Crash Space, where everyone would come in and share their work." Tuesday became his favorite day of the week. Next he created a robot spider that sat on his shoulder. "He was 85% hot glue and it kept falling apart."

Then came Dexter, a robot that was going to be more representative of him as a person. "I went through a whole process looking at different animals." It came down to choosing a monkey who was curious over a rabbit who was clever. His art school training kicked in. "I'm actually using a lot of my character design class ideas when I'm figuring out what the robot looks like."

Dexter satisfied the need that Jorvon once felt after sitting by himself in the college cafeteria. "That's what kind of started the whole robot craze. It was just me, saying 'Oh, okay. I'm by myself. I can fix that. I can build something.'" Dexter rides in Jorvon's backpack and peers over his shoulder. "Dexter is my personal familiar," he adds.

Now Moss has a job as a technician at a manufacturer of security systems in Culver City. "The funny thing is I got the job thanks to Crash Space," he explains. A lot of people who were working for this company were members of Crash Space. "One day I was walking by and one of my friends ran out to me: 'Hey, are you looking for a job?' I said, 'yeah.' So I ended up here." Even during Covid-19, he goes to the office, being considered an essential worker.

"Dexter is my personal favorite. And the thing about Dexter is that I'm always improving him. I'm always upgrading him. In a lot of ways, he's growing up and a lot of people see him as my child." Dexter is on V6, with over 160 components. He wants Dexter to become a wearable robot without the need of a backpack to carry various components. "I am designing a way for that to no longer be an issue," he says.

"Dexter is the robot I want people to see me with often," he says. At places like Maker Faire, when people see Dexter, they come up to talk. "A lot of the makers that I looked up to are my friends now, just because I kept making more and more stuff." The last Maker Faire Bay Area (2019) "is still the highlight of my life, the happiest weekend of my life." He adds: "I'm so wanting the Faires to come back because I miss it so much. I miss the people."

Jorvon Moss doesn't seem so odd anymore, at least not in the maker community. ✪

Follow @odd_jayy on Twitter and IG

See Jorvon's creations online at hackster.io/Odd_Jayy

Ball Drone
Mk II

Written and photographed by Benjamin Prescher

Build a unique single-rotor drone that steers with air vanes

The Ball Drone Project Mk II is a great way to get into single-rotor drone experimentation. It's an extraordinary drone design, totally different from the widespread multirotor aircraft — lets you really think outside the box.

This project shows a complete 3D-designed and -printed single-rotor drone. The self-made, 7"-round aircraft flies with a single rotor and steers with thrust-vectoring vanes. It uses common R/C components like radio receiver, LiPo battery, ESC, brushless motor, propeller, and servos. In addition, the commercial flight controller runs Betaflight, a popular free software option for drone racing.

Take a closer look at the four thrust vectoring vanes to fully understand how this thing is working. Four mini servos twist these vanes to direct the thrust, totally unlike your normal multi-rotor drone. It's a great learning experience and will surely generate excitement when you show it off to noobs and experienced drone pilots alike (Figure ).

I first published this project on Hackaday.io (hackaday.io/project/175512), as a complete remake of my earlier ball-drone project (/169823). The first version proved the concept, but I made it just too hard, building and programming my own controller from scratch. This new version is still a somewhat advanced project — skills (and tools) are needed in 3D printing, soldering, programming, and even drone flying — but this time it uses off-the-shelf R/C components and proven flight software that's free.

Anyone can build this Ball Drone. Learning to fly it is a whole new challenge!

TIME REQUIRED:
A Weekend

DIFFICULTY:
Intermediate/Advanced

COST:
$80–$100

MATERIALS
» **3D printed parts** Download the 3D files for free at thingiverse.com/thing:4635873.
» **Electronic speed controller (ESC), 32 bit, 35A** I used the T-Motor F35A 3-6S BLHeli_32, designed for FPV drone racing, Banggood #1221106, banggood.com.
» **Brushless motor, 2700kV, clockwise** Racerstar BR2306S, Banggood #1149532
» **Propeller, 6", 3-blade clockwise** FCMODEL 6045, Banggood #1006366
» **R/C receiver, 8 channel** to match your favorite transmitter. I used FlySky FS-A8S, Banggood #1092861.
» **Flight controller, Betaflight compatible** I used a Diatone Mamba F405 MK2, Banggood #1345007.
» **Mini servomotors (4)** Emax ES9051, Banggood #1039587
» **LED light strip (optional)** I used a strip of WS2812B RGB LEDs.
» **LiPo battery, 1300mAh (3 cells)** small and light. I used a Tattu 11.1V with XT60 connector.
» **XT60 socket** or whatever matches your battery
» **M3 hardware: 10mm screws (10), 12mm screws (4), nuts (8), and 15mm spacers (4)**
» **Cable ties**

TOOLS
» **3D printer**
» **Soldering iron and solder**
» **Screwdrivers / nut drivers**
» **Computer with Betaflight Configurator software** open source, free from github.com/betaflight

A

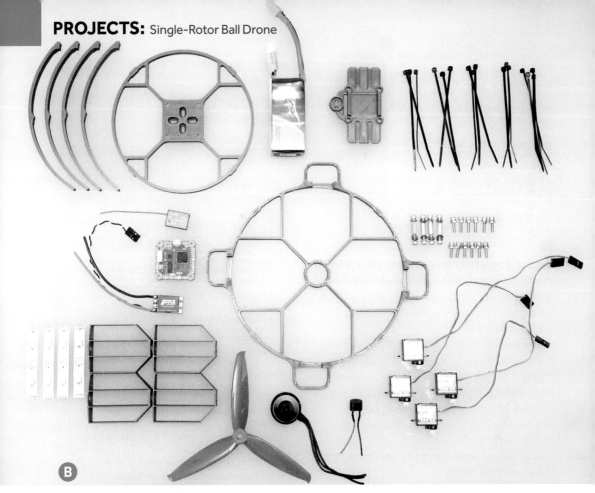

B

C

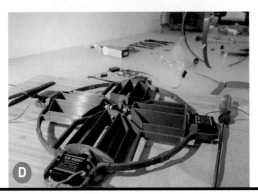

D

BUILD YOUR BALL DRONE

1. GET YOUR PARTS READY

3D print all the parts from thingiverse.com/
thing:4635873 (Figure B) and clean them up:
maybe you had some stringing, or maybe you
used supports for the drill holes. I printed the
drone in PLA, but I'd use PETG in the future
because of its strength. The construction is
designed for a layer height of 0.2mm and a nozzle
diameter of 0.4mm. Support is not required.

2. ADD THE SERVOS TO THE LOWER RING

Mount each servo in its bracket in the lower ring,
so that its axis of rotation is aligned with the
center of the drone.

3. CENTER THE SERVOS

Center each servomotor's rotor position by
sending it a 1.5ms pulse. For example, you can do

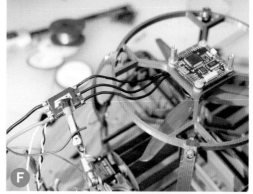

this with a simple Arduino sketch that calls the Servo library (arduino.cc/reference/en/libraries/servo). Once the servo is centered, mount the servo horn that comes with the ES9051 servomotors.

4. INSERT THE VANES AND BOLTS

Assemble one vane at a time. The servo horn fits exactly into the fitting on the vane wing. I used M3×12mm screws to make a positioning screw on the opposite side (Figure **C**). When all 4 wings are mounted, the lower ring is ready and can be put aside for a moment (Figure **D**).

5. ADD THE MOTOR ON THE UPPER RING

Mount the brushless motor first. Add screws for the flight controller from underneath so that the flight controller can be pushed down over the threads (Figure **E**). Then fix the flight controller in place using M3 circuit board spacers.

6. ADD THE LEGS

Add one drone leg after the other using M3×10mm screws with washers and nuts on the inside. Your drone should now stand on its own feet and look very similar to the final design.

7. CONNECT THE ELECTRONICS

Now it's a good time to solder up the ESC as well as the R/C receiver (Figure **F**). Depending on your battery connection, you can also solder "an XT60 plug to the supply from the ESC, as well as two small cables to supply your flight controller. I attached (daisy-chained) WS2812B lights on the inside of the drone legs and used these to connect the servo power lines (Figure **G**). The signal lines of the servos must be extended to reach the flight controller in the middle of the drone.

Next, solder the LED strip, the motor, and the

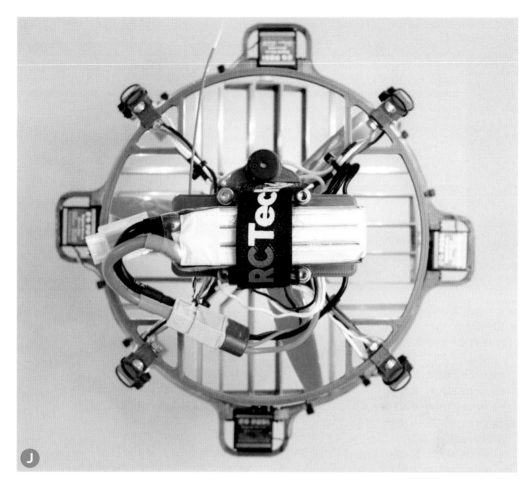

servo lines to your flight controller (Figure **H**). If your flight controller does not have any direct servo outputs, it is certainly possible to relocate these to the current motor connections via a "resource remapping" in Betaflight (see Step 8).

After you've soldered all the cables to your flight controller, you can screw the battery holder onto the spacer, and thread a battery strap through (Figures **I** and **J**). Assembly is done!

8. CONFIGURE THE BETAFLIGHT SOFTWARE

Contrary to my previous ball drone, this time I wanted to use a flight controller and software that are freely accessible to virtually everyone. There are a lot of open source projects, but Betaflight is currently the standard for racing drones (github. com/betaflight/betaflight). There are many tutorials and resources on the internet so I won't

explain Betaflight here. What we need to know is: How do you get from a racing quad to a single-rotor drone? Here's how:

• **RESOURCE MAPPING:** To control your single-rotor copter, you need one motor and four servos. Most flight controllers can run four motors and some of them have servo outputs as well. So what I did is remap my servo controls to Betaflight's standard four motor outputs. But then I needed a connection for my motor. In the configuration for the Fury F4 flight controller I used (github. com/betaflight/betaflight/tree/master/src/main/ target/FURYF4), I was able to see at which pins' *timers* are available. You often read about using the `LED_strip` port for a motor, but I wanted to use the `LED_strip` port for LEDs, so I decided to remap the PPM input pin as my motor output. To learn more about resource remapping and how to

do it, watch Joshua Bardwell's video at youtu.be/gL1DxUFjFq8.

The CLI commands I used:

```
resource MOTOR 1 NONE
resource MOTOR 2 NONE
resource MOTOR 3 NONE
resource MOTOR 4 NONE
resource PPM1 NONE
resource SERVO 1 A03
resource SERVO 2 B01
resource SERVO 3 B00
resource SERVO 4 A02
resource MOTOR 1 C09
save
```

- **MIXER CONFIGURATION:** Next, the flight controller must send the correct signals to the motor and servos. For this, you'll adjust the *mixer*. Under the Configuration tab, select Custom Airplane in the Mixer section (Figure K). Handle this via the command line:

```
# smix script for singlecopter on
MambaF405_MK2 (by Benjamin Prescher)
mixer CUSTOMAIRPLANE

# load a standard motor mix
mmix reset
mmix load airplane # Motor1 as ESC output
# mmix 0  1.000  0.000  0.000  0.000
```

```
# smix
smix reset
smix 0 3 0  100 0 0 100 0
smix 1 2 0 -100 0 0 100 0
smix 2 4 1  100 0 0 100 0
smix 3 5 1 -100 0 0 100 0
smix 4 3 2 50 0 0 100 0
smix 5 2 2 50 0 0 100 0
smix 6 4 2 50 0 0 100 0
smix 7 5 2 50 0 0 100 0
save
```

If you want to find out more about what's actually happening here, take a look at the Betaflight docs at github.com/martinbudden/betaflight/blob/master/docs/Mixer.md.

9. STABILIZE THE DRONE WITH SERVO PIDS

As you can see in Figure K, I have set a relatively low PID loop rate of 1kHz. The servos I'm using can usually be operated with a maximum **servo_pwm_rate** about 333Hz (there are also servos that can certainly handle more). I've set my **servo_pwm_rate** to 250Hz, which corresponds to one-quarter of the PID loop rate. As far as I understand the algorithms in Betaflight, there's no point in setting the PID rate artificially high if the command for the actuators can only be updated at a fraction of that rate anyway.

I use strong P values (Figure L), which tend

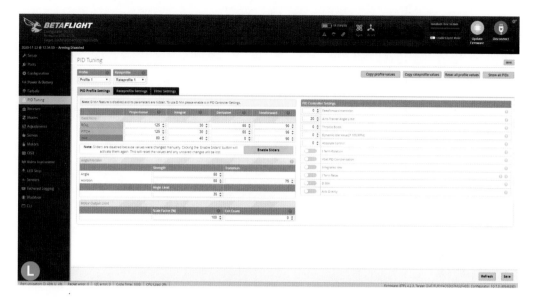

to make the vane servos "jitter." Betaflight has a great feature to solve the problem — a servo low pass filter. Handle this via the command line:

```
set servo_lowpass_hz = 20
set servo_pwm_rate = 250
save
```

For more on tweaking the PIDs, be sure to read my latest logs at Hackaday.io.

SOLO FLIGHT

Once you've set up the drone, connected everything, and implemented the configuration, test it out. Your drone should behave as follows:

- **Transmitter roll right** — forward and rear vanes move right
- **Transmitter pitch forward** — left and right vanes move forward
- **Transmitter yaw right** — forward vane moves left, right vane moves forward, rear vane moves right, left vane moves back

You're now flying a unique single-rotor, thrust-vectored drone that's ripe for experimenting. I would love to see what you do with it.

CONTROLLER OPTIONS

Betaflight is just one option I used to get this drone in the air. All of the above steps can be done using iNav as well (though I haven't tested it), and

I've heard that Ardupilot should do the job too! Try using a GPS with iNav for more flight features that aren't supported by Betaflight for now (such as altitude and position hold). Have fun!

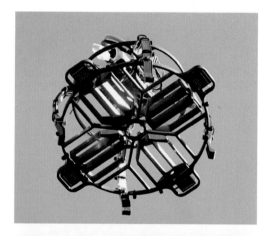

BENJAMIN PRESCHER is a passionate embedded systems developer and software engineer in Munich, Germany. He holds a master's from Technical University of Ilmenau in electrical power and control engineering, is deep into brushless motor control applications, and likes to look outside the box. The "drone thing" is just one of his hobbies.

Leave a like or follow at hackaday.io/project/175512, and share your makes at thingiverse.com.

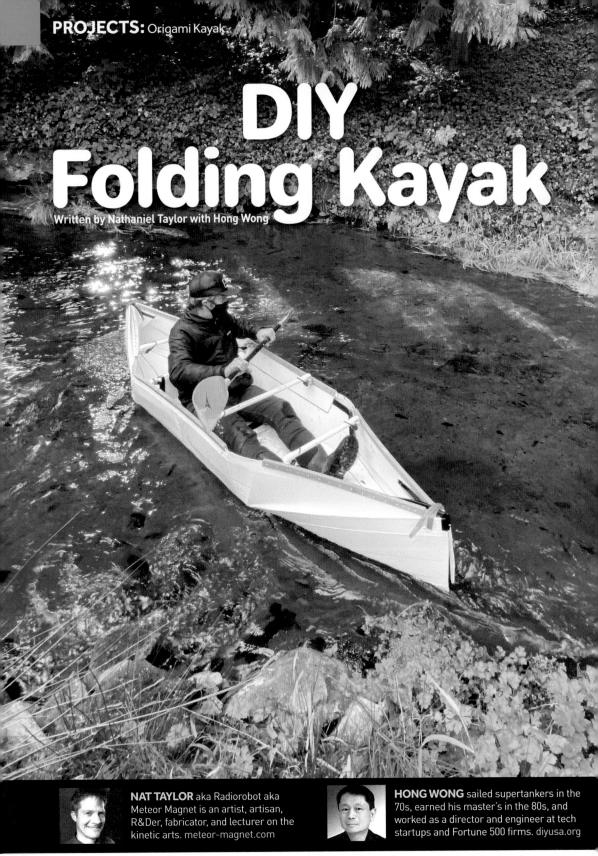

DIY
Folding Kayak

Written by Nathaniel Taylor with Hong Wong

NAT TAYLOR aka Radiorobot aka Meteor Magnet is an artist, artisan, R&Der, fabricator, and lecturer on the kinetic arts. meteor-magnet.com

HONG WONG sailed supertankers in the 70s, earned his master's in the 80s, and worked as a director and engineer at tech startups and Fortune 500 firms. diyusa.org

TIME REQUIRED A Weekend
DIFFICULTY Intermediate
COST $140–$200

MATERIALS

- » **Corrugated twin-wall polypropylene sheets, 4mm–6mm thick, 4'×8' (3 or 4)** depending on your weight. You know it as plastic yard sign material. I used 4mm Coroplast from the big box store, but thicker is better! Sign companies like Glantz can deliver 5mm and 6mm in many areas. If you can find sheets 147" or longer (good luck!) then you won't have to fuse them together.
- » **Vinyl lattice cap moulding, 8' lengths (2)** any color, Home Depot 207041505 or Lowes 602653
- » **PVC pipe, Schedule 40, ¾" dia., 10' length**
- » **PVC street elbow fittings, ¾" (8)**
- » **Bungee cords, 24" (4)** with soft or plastic end loops. Avoid sharp metal ends that will poke the kayak material. You can also use straps instead.
- » **100% silicone caulk, clear, 1 large tube**
- » **Gorilla Permanent All-Weather Tape, black** Amazon B07GRJ8L55 or similar
- » **Gorilla Waterproof Patch & Seal Tape, white** Amazon B08126M2R4 or similar
- » **Pipe clamps, galvanized, 1" (8)**
- » **Screws, ideally 18mm (100+)** for gunwales. I used 5mm stainless button cap screws 16mm long. 20mm will poke through. Or use wood screws — mind their sharp points — and you won't have to tap the holes.
- » **Machine screws, button head, 5mm×22mm, stainless steel (16) with nuts and lock washers (or lock nuts)** for bolting pipe clamps
- » **Voile straps, 15" (2)** any strap that can be tied taut
- » **Parchment or baking paper, 25' total length**
- » **Wheel, 4" or so, solid rubber** or similar. Needs a nice round profle to form the folds. Hong uses roller blade skate wheels; I carved down a 4" dolly wheel.
- » **Shaft, 12", with 2 fitting collars** to fit your wheel

TOOLS
Tools depend on material choices; here's what I used.
- » **Clothes iron or heat gun** The hot iron works best!
- » **Mat knife or box knife** for cutting Coroplast
- » **Tape measure, 12'–25'**
- » **T-square, drywall ruler, or straightedge** 48" ideal
- » **Clamps (optional)**
- » **Cordless drill** to tap and drill the vinyl lattice cap
- » **Gun tap, 5mm** aka chip-clearing tap
- » **Drill bit, 0.1963", #19, or 4.2mm** to start tap holes
- » **Soldering iron**
- » **Caulking gun**
- » **Safety gear**
- » **Permanent marker**
- » **Flat boards (2)** to hold sheets down while fusing
- » **Hex key set, metric**
- » **Adjustable wrench, small**
- » **Heavy duty snips** if you want to trim something
- » **Hacksaw** for cutting vinyl channel and PVC pipe
- » **Sandpaper, 120 grit** just a tad bit
- » **Rasp, grinding wheel, or belt sander** if you're shaping your own roller wheel

Escape from it all in this lightweight, foldable boat made of ordinary corrugated plastic

Out of the blue sky — or was it wildfire gray at the time? — I received a call to build a crazy origami folding kayak designed by Hong Wong. Like any hungry, cooped-up human who loves boats and the river, I seized the moment! If you like boats and boat building, you will appreciate the economy of materials, tools, and engineering in this boat.

Over two weeks, I built a few versions, trying as many melting, hot iron-fusing, and dim-witted thermoforming techniques as I could think of. In the end, the path was simple — a clothes iron is all that's needed to bond the plastic sheets. The final boat was built in two days and there's a time-lapse video to prove it! (You can watch it on the project page at makezine.com/go/folding-kayak.)

This 12-foot, corrugated plastic origami boat is a low-cost build for those of us who like to experiment on the water or land. It paddles great and costs a fraction of the store-bought variety. And it's just a cool way to build with plastic sheet. The folding, origami-like transformation from flat to boat to refolded for transport is awesome.

Hong has other versions of his boats and even one for low-cost disaster relief. Whatever, man! It's just good to get outside. Let's build a folding plastic kayak!

> **DISCLAIMER:** This is an experimental boat. We cannot guarantee your safety. You assume all the risks of building and using this boat. Read the liability waiver at makezine.com/go/folding-kayak.

BUILD YOUR ORIGAMI KAYAK

Before you build, download the free PDF plans (Figure Ⓐ on the following page) from the project page makezine.com/go/folding-kayak and watch the folding/unfolding video there too.

Nathaniel Taylor/Mariano Di'Yorio

Gen 14 Foldable Kayak. ©: Hong W. Wong, 2020. Plan version: Make: Magazine. **Foldable Can'Yak.** Nat Taylor, 2020.

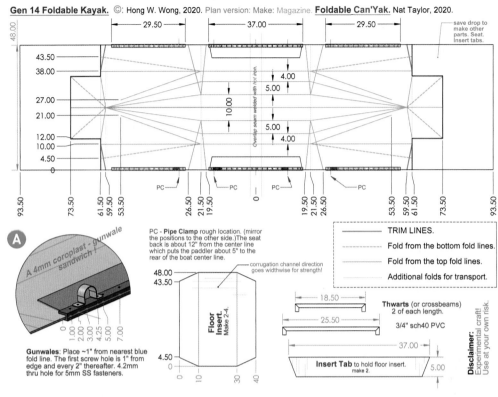

save drop to make other parts. Seat. Insert tabs.

Overlap seam welded with hot iron.

corrugation channel direction
goes widthwise for strength!

PC - **Pipe Clamp** rough location. (mirror the positions to the other side.) The seat back is about 12" from the center line which puts the paddler about 5" to the rear of the boat center line.

TRIM LINES.
Fold from the bottom fold lines.
Fold from the top fold lines.
Additional folds for transport.

Floor Insert. Make 24.

Thwarts (or crossbeams) 2 of each length.
3/4" sch40 PVC

18.50
25.50
37.00

Insert Tab to hold floor insert. make 2.

Disclaimer: Experimental craft! Use at your own risk.

A 4mm coroplast - gunwale sandwich!

Gunwales: Place ~1" from nearest blue fold line. The first screw hole is 1" from edge and every 2" thereafter. 4.2mm thru hole for 5mm SS fasteners.

1. FUSE TWO PLASTIC SHEETS

Fuse two 4'×8' sheets of the corrugated polypropylene together using a hot clothes iron set to high, with a large piece of parchment paper protecting the sheet always. Align the sheets with an overlap equal to your iron width. Go slowly and move the iron back and forth over the overlap (Figure **B**). Take care to heat the sheet slightly wider than needed, in order to create a rolling edge to the overlap. Press down and wait a bit (Figure **C**). If there's a mistake, patch it with more Coroplast!

Coroplast melts at 375°F. Let it soak up the heat into the lower layers so that all the layers fuse together. Use the lumber to hold the sheets flat and to rest your hands on while working over the sheets.

Flip the sheets and fuse them on the other side too. All edges of the overlaps fused? No huge holes in the boat? You're ready to move on (Figure **D**).

2. LAY OUT THE PLAN

The plans are drawn for easy layout. Use marking and measuring tools to draw the plan onto the

BOAT DREAMS

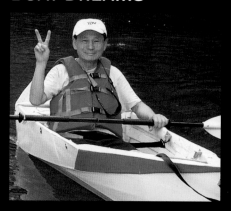

Hong W. Wong is an inventor at heart. A retired Intel engineer with 100+ patents, he shares plans and tutorials at diyusa.org for making musical instruments, paper crafts, and especially, Coroplast boats. When he presented a new folding boat at Virtually Maker Faire in 2020, we had to know more.
—*Keith Hammond*

Why did you start building these boats?
I came from a poor family. As a kid, I lived near the sea and I wanted a boat but never had money for one. Now I can afford a boat, but I don't have space to keep one! Inflatable boats need more maintenance, and foldable kayaks are expensive, so I started learning how to design and build foldable boats. I want to share my learning so others can enjoy the fun of kayaking without spending a lot of money and time.

Does the folding affect the boat strength?
Yes, so we must add structural elements to compensate. This can be a cross-beam made with PVC pipe, or double layers of Coroplast on the floor, or a properly designed seat. There are other tricks too.

Which Coroplast is best for folding boats?
I use 6mm thickness, and there are 5mm boats on the market, but many of my YouTube viewers asked how to build with 4mm sheets from Home Depot/Lowes. Last summer I discovered the heat fusion method and I'm glad Nathaniel improved it!

Nathaniel Taylor/Mariano Di Yorio, Yannie Wong

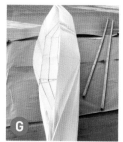

two fused sheets (Figure **E**). Start with the midsection line at the best location for the boat centerline. The boat plan lines are mirrored; measure equally on all sides, front to back.

Pay attention to the front versus back fold lines on the plans. To project the back fold lines to the other side, hold the sheets up to the light (Figure **F**). Clearly mark the cut lines.

> **TIP:** Try printing out the plans and folding a paper model first — it really brings some understanding of how the folds work (Figure **G**).

3. CUT OUT THE BOAT
Cut out the entire flat shape, following the red trim lines (Figure **H**). Keep all the waste; you'll use this to form a seat and some of the rail assembly.

4. BUILD THE ROLLER
This is kind of like a rolling pin for creating fold lines in the sheet. The wheel is shaped so that it

presses a wide deformation into the corrugated sheet with less friction than other folding tools. If it's too narrow, the pointed load will tear the sheet. Find a solid rubber or similar wheel that can shaped into an ovaloid shape as shown (Figure ⓘ). A rollerblade or dolly wheel can work. Insert a shaft through the wheel and trap the roller midpoint on the shaft with two shaft collars. Make sure it can roll smoothly.

5. ROLL THE FOLDS

Drink some water! Work out those shut-in abs and get rolling. Roll along the lines to create creases. Roll everything that needs rolling. Roll it again, front and back (Figure ⓙ).

> **TIP:** I used brute force on an indoor floor, but Hong has his own technique: he uses a heat gun to soften the fold line and then rolls the fold over a concrete joint in his driveway to help shape it. Check his videos at diyusa.org.

6. FOLD THE SHEET

Fold the plastic sheet as you roll the lines, front and back, to create the valley and mountain folds. Wrestle the sheet until the kayak takes shape (Figure ⓚ). Welcome to the second part of the staycation workout plan. *Coroplast is forgiving.* Continue to fold and shape the kayak along the plan lines before the gunwales go on.

> **TIP:** Use clamps to hold the boat together while folding. And use chunks of stiff cardboard to protect the boat while you're wrestling.

It's probably the end of day one, or about halfway through the project. Take a break and start daydreaming of a safe test-pond location.

7. CUT THE STRAP SLOT

Once the boat can fold into its basic shape, create a slot on both ends though all the front-folds to "sew" a strap through and bind the folds together in boat form (Figures ⓛ and ⓜ). Make this slot

slightly larger than the width of your strap.

With the boat folded, drill two through-holes, so that when unfolded, these holes make registration marks for the slots. Unfold the boat, cut each slot with a mat knife connecting the two-hole pattern, and open the slot by melting it with a hot soldering iron.

8. MAKE THE GUNWALES

Cut the gunwales from the plastic lattice cap: the midsection gunwale is 25½" long, and the front and back gunwales are equal length, made from what's left over. Then mark the screw holes every 2", beginning and ending 1" from each end (Figure **N**). While you're at it, mark the holes for bolting on the pipe clamps as shown in the plans.

Pre-drill all the holes now. This will help with alignment of the thread tap, and screws, passing through the rail-and-Coroplast sandwich later.

9. MAKE THE INSERT TABS

Cut the two insert tabs from one of the waste sections of Coroplast, with the vertical corrugation following the short dimension. Their sole purpose is to hold the folded floor insert that adds rigidity to the boat.

10. MOUNT THE MIDSECTION GUNWALES

Take the midsection gunwale and sandwich the insert tab and the boat midsection into the lattice cap slot (Figure **O**). If necessary, use the hot iron to flatten the edge of the sheet, or use a carpet knife to carefully cut down the center of the sheet edge, cutting only the ribbing. Build a good sandwich.

Tap threads in one of the end holes. Insert the first 5mm×16mm screw to hold the gunwale in place. Insert your tap into a drill to speed this way up (Figure **P**)! Tap the threads quickly and do not bottom out the tap. To speed up driving the fasteners in, buy a proper hex bit or make one from a hex L-key.

Continue to tap and screw the 5mm×16mm screws into the midsection gunwale. Keep in mind any sharp points: Make sure there are none!

Repeat these steps to mount the other midsection gunwale.

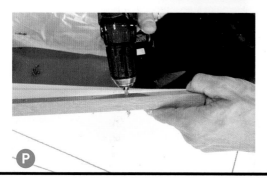

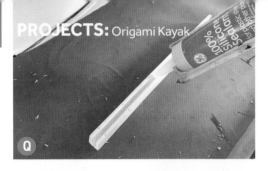
Q

R

S

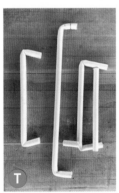
T

U

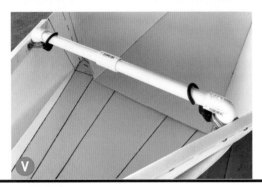
V

11. MOUNT THE FRONT AND BACK GUNWALES

Fill the slot in the lattice cap with 100% silicone, then press the silicone-filled slot over the boat edge (Figures Q and R). Tap threads for the machine screws in place, right through the silicone and the boat sandwich, using the predrilled holes as guides. Again, you can put your tap and hex bits in a power drill to go faster.

12. MOUNT THE PIPE CLAMPS

Bolt each pipe clamp in place at the locations you marked in Step 8. Make sure you use lock washers or lock nuts here (Figure S).

13. BUILD THE THWARTS

Use ¾" PVC pipe and create 90°-bent ends with the correct width as marked on the plan. There are a few ways to do this. Heat-bending the pipe is one reliable way (Figure T). Creating an assembly out of plumbing parts (elbow fittings) is another (Figure U). Both work well. An optional version, as tested, is to build thwarts like tent poles with elastic hook ends (Figure V). Thwarts should be built for rigidity.

14. BUILD THE FOLDED INSERT

This cross-brace stiffens the boat. Follow the layout pattern to orient the sheet in the correct direction, then cut the patterns. Fold and tape 2 patterns together, or use an extra sheet to make 4 for more support (Figure W).

15. BUILD THE SEAT

With the remaining material, build a seat (Figure X). Use black Gorilla tape to hold it all together.

16. SEAL THE CENTER SEAMS

Your fused center seams may look watertight, but let's make sure. Tape the center seams with the white Gorilla weather seal tape. Do this on the outside and inside, making sure to overlap your fused seam on both sides (the two opposing tapes will be offset). This will waterproof your seam.

> **TIP:** Find extra hands to help with the taping, as this tape is super sticky.

17. TAPE ANY TROUBLE SPOTS

Use the black Gorilla tape over all exposed open cells and any sharp points. Tape the bottom fold line intersections too, as these will wear the most. Tape whatever's good for your boat (Figure **Y**).

18. PATCH ANY LEAKS

Check the boat for leaks or punctures by holding it up to the sun and looking for light leaks. Patch these with the weather seal tape or by heat-fusing a small patch of Coroplast over the hole.

You're ready to water test (Figure **Z**)! Good luck. Be safe and wear a life vest.

FOLD AND FLOAT!

This kayak is fun to build. It folds into a form that you can transport in a car trunk and refolds into a boat in about 3 minutes (Figure **Aa**). That's fun too! And of course it's fun to paddle around.

Keep your kayak clean, store it indoors, add paint if you like. Avoid sharp hazards and shorelines.

> **NOTE:** I call my boat a "Can'Yak" because it's got thwarts but no deck. For rougher waters, there's a deck too; check Hong's Gen 12 plans at diyusa.org.

I would have enjoyed thicker material for the Coroplast. I used what I could find. The plastic corrugated kayak industry uses 5mm. Hong uses 6mm! I was on the weak side with the 4mm sheet, and as I paddled along, one passerby yelled, "Nice coffin!" As an experimental boat, not so far from the truth, yet, I disagree. On the second trip, I added an additional folded insert at an offset, and I felt confident I could paddle on a calm lake or pond, so I made a day of it. But the thicker material could lower the fear of puncturing the boat.

Now that I have a few hulls, my thoughts are drifting toward experimenting with different ideas and modifications. Yesterday I figured out a way to make a shelter out of two of them (Figure **Bb**).

All in all, a fun project for those of us that like to get wet and tinker in the water. Build a boat by spring and enjoy some outdoor water adventures when the weather warms up again. Yahoo! ●

Special thanks to SkanlanKemperBard (skbcos.com) for the use of their amazing loft!

W

X

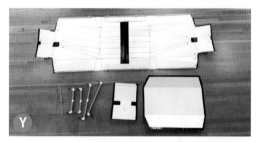

Y

Z

Aa

Bb

Nathaniel Taylor/Mariano Di Yorio

Live Streaming With Raspberry Pi

Set up and stream a remote viewable video camera, using everyone's favorite micro computer

Written and photographed by Eben Kouao

Real-time video streaming is a great way to interact with our environment, share what we're making or teaching, or just keep an eye on things, like a CCTV camera. But what is the technology behind it?

And what if we could do it with a Raspberry Pi — by streaming the Pi Camera's live feed to any client devices we allow?

In this tutorial we're going to do exactly that. We'll show you how to build a live camera that streams to your phone or other device. If you

recently got a Raspberry Pi and want to work on a new project, this is a good one.

By the end, you'll be able to set up a wireless stream to view the camera feed from any device on your network (Figure Ⓐ). This means you can create your own mini camera stream for any application — think houseplant monitor, security CCTV camera, or even a portable streaming cam. We'll cover the assembly, installing the OS, and setting up the stream in Python. So let's get started!

TIME REQUIRED:
1–3 Hours

DIFFICULTY:
Intermediate

COST:
$50–$70

MATERIALS

» **Raspberry Pi 4 single-board computer, 4GB RAM or more** Get it in our Getting Started with Raspberry Pi Kit from the Maker Shed, makershed.com, or buy separately from other resellers.
» **MicroSD card, 32GB with adapter**
» **Raspberry Pi Camera Module v2** With 8MP resolution, it takes 1080p HD video, rpf.io/camv2yt. You could also use the Raspberry Pi High Quality Camera, rpf.io/hqcam, more expensive but at 12.3MP a big improvement in image quality. For this project, we're using the cheaper Camera Module.
» **Micro HDMI to standard HDMI cable** or use a regular HDMI cable with our micro HDMI adapter, in the Maker Shed kit
» **Raspberry Pi Power Supply, USB-C** also in our Maker Shed kit
» **Pi enclosure (optional)** also in Maker Shed kit
» **Ethernet cable (optional)**

TOOLS

» **Computer with internet access and SD card slot**
» **Keyboard and mouse** for setup only; not needed afterward
» **Monitor (optional)** if you'd like a connected display, in addition to web streaming

EBEN KOUAO makes DIY projects and prototypes, from smart mirrors to electric skateboards and more. Watch his tutorials at youtube.com/ebenkouao and youtube.com/make.

MEET THE RASPBERRY PI 4

The marvelous Raspberry Pi (Figure Ⓑ) is a complete computer — a very low-cost computer. Like your MacBook or desktop PC, it features USB ports for peripherals and other connectors such as an audio jack, Ethernet, and HDMI ports.

These boards are getting more powerful over time, and are now able to do nearly everything your regular computer can do. The interesting part is the form factor; because of its small size, this "credit card computer" is useful for many different real-world applications and it's popular in the DIY community.

For this project, we'll be using the Raspberry Pi 4, the newest and most powerful model. You can use other models, but performance may vary.

WHAT IS STREAMING AND HOW DOES IT WORK?

We're capturing live footage from the Raspberry Pi camera and using Flask software to create the live stream to our client devices — any device that can access the web (Figure Ⓒ). Flask can be seen as a back-end web server and micro-framework for Python, making it easier to build web frameworks using Python. It's a great bridge between Python and HTML web pages, and it supports Motion JPEG (M-JPEG or MJPEG), a

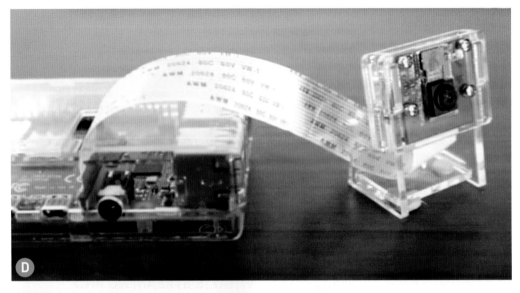

D

E

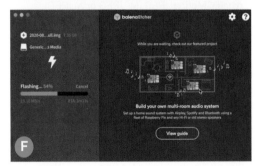

F

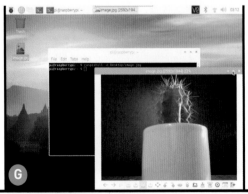

G

video format that works by streaming a sequence of independent JPEG images. Motion JPEG is widely used in security IP cameras and webcams alike, and this is how we'll stream our live feed.

Further, a device connected to the same network as the Pi can also visit the URL to see the live footage, using the Flask framework for video streaming.

BUILD YOUR RASPBERRY PI LIVE STREAMING CAMERA

CONNECTING THE HARDWARE
1. CONNECT THE CAMERA MODULE
The Raspberry Pi Camera Module provides the features to take pictures, record videos, and stream video.

While the Pi is powered off, lightly pull on the Camera port latches to expose the camera module port. Insert the camera ribbon cable and push the latch back into position. Typically the correct orientation is with the blue part of the ribbon cable facing toward the USB ports.

NOTE: Handle the camera port clips/latches with care as they're quite fragile.

2. CONNECT AN ETHERNET CABLE (OPTIONAL)
If you plan to use a wired network connection for your stream, plug your Ethernet cable into the Pi.

3. CONNECT THE HDMI OUTPUT

If you choose to use a monitor, connect an HDMI cable to the port of your TV or monitor, and to the Raspberry Pi's Micro-HDMI output port. You'll need a Micro-HDMI cable or adapter for this.

And that's the whole hardware installation (Figure **D**).

CONFIGURING YOUR RASPBERRY PI
4. ETCH THE PI OS TO THE MICROSD CARD

Before you can turn this Pi on, you'll need to install an operating system onto the microSD card. We're using the standard Raspberry Pi OS (Figure **E**), formerly called Raspbian; you can download it from rptl.io/raspberrypios.

Once you've downloaded the image, flash it onto the microSD card using Balena Etcher (balena.io/etcher). This process can take up to 15 minutes, depending on your SD card type (Figure **F**). Once completed, you can safely eject the microSD card.

Now insert the microSD card into the Raspberry Pi 4 and power up the Pi by plugging in the USB-C cable from the power supply.

> **NOTE:** I wrote this project using the August 2020 release of RasPiOS, there's a newer one now.

5. SET UP THE PI OS

First make sure your Pi has an internet connection via Wi-Fi or Ethernet:

- **Wired connection** — Plug the Ethernet cable from the Pi into the LAN port of your router.
- **Wireless** — You can search for your Wi-Fi Access Point and enter your credentials to log in wirelessly.

Once logged into the Pi, complete the setup installation dialog boxes and finish.

Next you'll enable the camera port and VNC server. The camera port enables the camera to be used on the Pi. VNC (virtual network computing) lets you access the Pi without the need for a display, so you can access your video stream remotely without hooking the Pi up to a monitor again.

Open the terminal window and run:

```
sudo rasp-config
```

On the configuration menu, select Interface Options → Camera Port, and select Enabled. Then select Interface Options → VNC, and select Enabled. Optionally, enable SSH to access the Pi via your laptop. You can now exit. This may cause your Pi to reboot.

6. TEST YOUR CAMERA SETUP

Once you reboot you Pi, confirm that your camera works. Take a picture by entering this command into your terminal:

```
raspistill -o Desktop/image.jpg
```

This command will take a picture and save it to the desktop of your Pi (Figure **G**). You can also record a little a video output to test that too:

```
raspivid -o Desktop/video.h264
```

SETTING UP A PYTHON CAMERA STREAM
7. INSTALL DEPENDENCIES

Now that you've confirmed the camera module is working you can move into installing the following dependencies and the GitHub repo.

First update your Pi to make sure all your libraries and packages are up to date by running the following commands:

```
sudo apt-get update
sudo apt-get upgrade
```

> **NOTE:** It's good practice to create a virtual environment for the Python3 library dependencies. The reason is that you want to install Flask in a container within the Raspberry Pi, so that if anything goes wrong, you don't need to start again. However, since we're installing only a few libraries here, we'll skip this.

Python 3 is already installed in RasPi OS. But depending on your Raspberry Pi you may need to install the following dependencies to create a live stream:

```
sudo apt-get install libatlas-base-dev
sudo apt-get install libjasper-dev
sudo apt-get install libqtgui4
sudo apt-get install libqt4-test
sudo apt-get install libhdf5-dev
sudo pip3 install Flask
sudo pip3 install numpy
sudo pip3 install opencv-contrib-python
sudo pip3 install imutils
sudo pip3 install opencv-python
```

Upon successful installation (Figure **H**), you're all set up to download the Github repo.

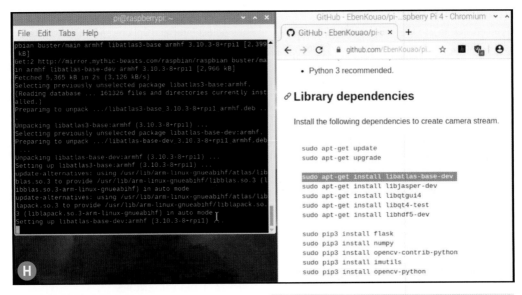

8. CLONE THE GITHUB REPO

Open terminal and clone the Camera Stream repo:

```
cd /home/pi git clone https://github.
com/EbenKouao/pi-camera-stream-flask.
git
```

A couple things to note:

- **Templates folder** — This is where the *index.html* webpage will be stored. This basic HTML site is what the client will view in the browser, and is fully customizable. The **<img>** image element is constantly updated by the browser with the latest stream of JPEG images.
- **Python scripts** — The *Camera.py* script accesses OpenCV and enables the camera module's output, providing the sequence of frames via Motion JPEG. The *main.py* script is where the Flask stream is created (Figure **I**). The main application imports a **camera** class (module).

9. LAUNCH YOUR WEB STREAM

Now you can start the Flask camera stream with the following command:

```
main.py ×
 1  #Modified by smartbuilds.io
 2  #Date: 27.09.20
 3  #Desc: This web application serves a motion JPEG stream
 4  # main.py
 5  # import the necessary packages
 6  from flask import Flask, render_template, Response, request
 7  from camera import VideoCamera
 8  import time
 9  import threading
10  import os
11
12  pi_camera = VideoCamera(flip=False) # flip pi camera if upside down.
13
14  # App Globals (do not edit)
15  app = Flask(__name__)
16
    @app.route('/')
    def index():
        return render_template('index.html') #you can customze index.html
```

```
sudo python3 /home/pi/pi-camera-
stream-flask/main.py
```

Where */home/pi/pi-camera-stream-flask/ main.py* is the direct path to your Python script. Alternatively, you can access the file directly and run the Python script that way.

10. MAKE THE STREAM AUTO-START (OPTIONAL)

It's a good idea to make the camera stream auto-start at bootup of your Pi. Then you won't need to re-run the script every time you want to create the stream. You can do this by going editing the */etc/profile*:

```
sudo nano /etc/profile
```

Go to the end of the file and add same terminal command from above:

```
sudo python3 /home/pi/pi-camera-
stream-flask/main.py
```

This causes the terminal command to auto-start each time the Raspberry Pi boots up. In effect, this creates a "headless" setup, which you can then access via SSH.

NOTE: Make sure SSH is enabled. If necessary, return to the **rasp-config** instructions in Step 5.

NOW STREAMING LIVE, FROM YOUR LIVING ROOM

It's time to tune in!

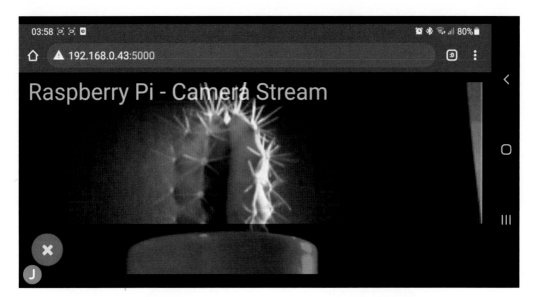

Let's demonstrate that Flask implementation. You can view the live stream from your Raspberry Pi by accessing its URL on port 5000. Just visit <your Pi's IP address>:5000 in a browser to access your video stream on any devices connected to the same Wi-Fi network as your Raspberry Pi (Figure **J**).

> **NOTE:** You can easily find out your Pi's IP address. Enter `ifconfig` into your terminal command, then look for the `Inet` address.

And that's all! Now you've created your own Raspberry Pi camera stream.

STREAM LATENCY

The Pi stream's latency depends on your network coverage; you can expect about 1–2 seconds latency, but in some cases as fast as ~500ms.

Since you're streaming over your local network, a few factors may affect the performance of your stream such as your Wi-Fi coverage, Wi-Fi performance, and your Raspberry Pi version (that's why the Pi 4 is recommended for best performance).

NEXT STEPS

Building your own Pi camera stream is a great start into the world of live streaming projects. Using a Raspberry Pi brings down the cost significantly while retaining a good amount of processing power.

This tutorial is a fun build to get you started, but there are lots of ways to take it further. What live-stream applications will you make next? Plant monitor? Portable DIY "GoPro" or body camera? Smart CCTV camera with face recognition?

And how might you take it further to view the stream remotely when you're not connected to your home Wi-Fi network? Say, from your computer at work? VPN or SSH tunneling?

Stay tuned for my next build! **⊘**

Video streaming with Flask — more information on the theory: blog. miguelgrinberg.com/post/video-streaming-with-flask

Weasley Whereabouts Clock

Written and photographed by Patrick Peters

Build a real location clock that knows and shows where your family and friends are

PAT PETERS is a software developer living in Council Bluffs, Iowa, with his wife, Christine, daughter, Eleanor, and two cats. In his free time, he teaches brass instrumental music lessons and programming classes at the 402 Arts Collective in the Omaha metro area. He also likes running, woodworking, and playing with his daughter.

Ever since I was 8, I've loved Harry Potter and the world that J.K. Rowling built with her books. I always related to Harry's friend in the book, Ron Weasley, and I especially loved that Rowling tied the wizarding world into the real world.

I wanted to bring a piece of that world into my own (while also geeking out with Raspberry Pi and GPS), which led me to re-create this famous prop from the book. This project is an Internet of Things Location Clock (or Whereabouts Clock, or Weasley Clock). Rather than having two hands that give you the time of day, this clock has a hand for each member of your family or group, and moves the hand to point to wherever that person is in the real world! My clock, for example, has two hands (for my wife and me) and shows locations for Home, School, my Parents' house, our favorite bar, the opera house (we volunteer there), etc.

The Location Clock runs on a Raspberry Pi mini computer, which subscribes to an MQTT broker — a server for routing messages in the MQTT internet protocol. Our phones send a message to the broker anytime we cross into or out of GPS waypoints we've set up in the OwnTracks phone app, which then triggers the Pi to run a servo that moves the clock hand to show our location. And the Node-RED programming tool made it easy to automate without writing any code.

I've seen a couple different approaches to this clock idea, but I especially valued those of Allie Fauer (instructables.com/IoT-Location-Sensing-Picture-Frame), and Alya Amarsy (instructables.com/Weasley-Clock). Here's how I made mine.

DESIGNING THE CLOCK

I wanted to give my Weasley Clock an authentic, old-fashioned look, like it's been in the family for a while. On Craigslist, I eventually found an old-fashioned mantel clock that had broken down and was beyond repair (Figure A).

The clock face was the one piece I knew exactly how I wanted to look. I admit, my Photoshop skills aren't quite up to snuff to make this from scratch, so I surfed the interwebs and found a design pattern I liked, from ginnylovegood1.weebly.com. I downloaded her JPEG (Figure B on the following page), imported it into Photoshop, and cut out the clock face. Next I had to find the

TIME REQUIRED:
A Few Weekends

DIFFICULTY:
Intermediate

COST:
$175–$200

MATERIALS
» **Old beat-up clock** or clock shell
» **Servomotors, 360° (2)** aka sail winch servos, GWS model S125 1T 2BB
» **Raspberry Pi single-board computer** I used a Raspberry Pi 1 model B+ but pretty much any Pi will do.
» **Adafruit PWM/Servo Pi-Hat** item #2327, adafruit.com. I had to solder the headers.
» **Clamping hubs (2)** to fit your servo shafts, servocity.com/standard-clamping-hubs
» **Spur gears, hub mount, 48 pitch: 96 tooth (2) and 66 tooth (2)** Two for the servos, two for the brass tubes that turn the clock hands. kimbroughracingproducts.com #142 and #301
» **Brass tubes (2)** from K&S Hobby. I used the ¼" and 9/32" diameter tubes, since they can fit into each other.
» **Machine screws, #10×1½" (2)** with nuts
» **AlumaMark laser markable aluminum (1 sheet)** gold or brass color
» **Plywood, ¼", 12"×12"**
» **Acrylic sheet, clear, ⅛" thick, scraps**
» **Rubber grommet, 1⅛"** for topmost clock hand

TOOLS
» **Laser cutter**
» **Computer with internet connection** for setup only; not needed for clock operation
» **Saw and drill**
» **Tinsnips**
» **Glue**
» **Soldering iron and solder**

A

Harry Potter-styled type font, so I could replace the destinations with what I wanted (Figure C). I converted my Photoshop document to a vector, then made an Adobe Illustrator vector to work with (Figure D).

I wanted the clock to look antique, so I decided to etch the design onto metal. You can take several different approaches here. The easiest was to etch the design with a CNC laser engraver. I originally bought a sheet of solid brass but the laser wasn't strong enough to etch it. Then I found an alternative: AlumaMark is thin, flexible aluminum sheeting that's designed for marking with lasers. I took my Illustrator file to the Omaha Do Space, and "etched" my design with their laser cutter (Figure E).

Lastly, I cut the face out from the metal sheet by hand — first a rough circle using a jigsaw, then a little more detail work with some tinsnips — and I had my clock face (Figure F).

I designed the clock hands in Photoshop (Figure G), converted them to a CNC-friendly vector graphic in Adobe Illustrator (Figure H), then laser-cut them out of ¼" birch plywood (Figure I). But I'd forgotten to add holes in the original design, and when I tried to drill them the brittle, scorched plywood broke apart. To work around this, I cut a bit of leftover acrylic to the same width, and drilled the hole through that. I then glued the broken clock hands onto the acrylic, and painted the whole assembly with black spray paint.

Finally, I printed out two small pictures of myself and my wife to attach to the clock hands (Figure J).

MOVING THE HANDS — HARDWARE

This part was by far the most challenging: How to get the individual clock hands to move freely of their own accord while remaining stacked on top of each other?

I spent a good deal of time looking at previous whereabouts clocks for inspiration. The majority used display screens to show clock hands, or light-up displays to show the different locations. I really wanted my clock to have moving hands, to give it a real, antique look.

I finally found a design, used by two other

B

C   D

E

F

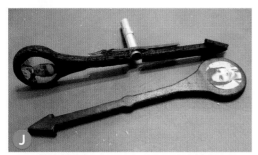

fellow magic clock makers, that relies on tubes of varying circumferences stacked inside each other and controlled by specialized sail-winch servos (used for R/C sailboats). The thing I really like about this design is that you can add as many hands as you like, until you run out of room for servos! I used two GWS model S125 1T 2BB servos that provide accurate, full 360° rotation (most servos only have a range of about 180°); I found them on AliExpress. You can find the K&S Hobby brass tubes, ⁹⁄₃₂" and ¼" diameter, at any hobby store, R/C store, or online.

I cut a mounting block for the servos and tubes out of a scrap of ½" plywood, and attached it to the clock's back door (Figure K). Each servo has a large gear mounted on its shaft; the top gear is connected to a spacer hub to stagger the two. The tubes fit into the hole in the middle of the block. On each tube is a smaller gear hooked up to a

screw-on gear hub (Figure L), so that when the servo turns, the tube rotates.

The Raspberry Pi I'm using has only one GPIO output pin that can supply a full 5 volts of output, but I need 5V to drive each servo. Thankfully, Adafruit makes a Servo Hat that can supply the additional outputs for the servos, and it fits right on top of the Pi (Figure M).

MOVING THE HANDS — SOFTWARE

The Adafruit Pi Hat comes with its own Python library for driving the servos. There's a really helpful reference at learn.adafruit.com/adafruit-16-channel-pwm-servo-hat-for-raspberry-pi.

You control servos with *pulse-width modulation (PWM)*. Without getting too much into the nitty gritty, PWMs are signals sent at a steady pulse in whatever frequency you need to control electronics like LED lights and servos. By

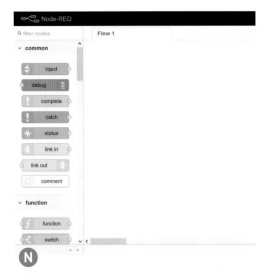

changing the length of time between the high and low signals, we can control how far the servo's shaft turns, which then points the clock hand wherever needed.

Using the Adafruit library, I set a bit of Python code for each hand position I need, then I call these from a *system command node* in Node-RED. The Python code treats the PWM instances as their own objects, so every time I need to set a hand's position, the code will:

- Create a new PWM object
- Set the exact frequency of the PWM object
- Set when in each pulse the signal should change from low to high and then back from high to low
- Close the PWM object, so the clock hand stays in whatever position I need.

On the project page at makezine.com/go/ weasley-clock, you can see a video of me running a test script and you can download my code for one of the hand movements, *location_clock.py*, based on the example code provided by Adafruit.

NODE-RED ON RASPBERRY PI

Node-RED is an open source Node.js programming tool that enables you to set up some fairly advanced programs and workflows without writing any code. It has an extensive library of "nodes" you can use to automate actions based on events such as an incoming text or email, a time of day, or (in our case) an incoming message from an MQTT broker.

Node-RED might already be installed on your Raspberry Pi OS, but older versions can be terribly out of date and insecure. To install the newest versions of Node.js, npm, and Node-RED all at once, type the following into your terminal on the Raspberry Pi and hit Enter:

```
bash <(curl -sL https://raw.
githubusercontent.com/node-red/linux-
installers/master/deb/update-nodejs-
and-nodered)
```

You can find more information at nodered.org/ docs/getting-started/raspberrypi.

Once you have Node-RED installed, start up the service by running the following command in the terminal:

```
node-red-start
```

Once the service is up and running, you can either go to 127.0.0.1:1880 in a browser on your Raspberry Pi, or if you're on the same internal network as your Pi, you can open a browser and type in the Pi's IP address on port 1880 to go to the Node-RED Setup screen. So, for example, if my Raspberry Pi has an internal IP address of 192.168.0.100, I would go to http://192.168.0.100:1880 to navigate to the setup screen (Figure N).

SETTING UP AN MQTT BROKER

MQTT (Message Queueing Telemetry Transport) is a popular messaging protocol that uses a publisher/subscriber model for sending machine-to-machine messages on spotty connections. The main thing I like about it is the amount of documentation out there about it, and that it is very lightweight on data usage. I have it running on my phone 24/7, and it barely uses 1 megabyte of my 4G data per month!

Before you can send data via MQTT, you need to set up an *MQTT broker* to route messages. An MQTT broker acts as the server that you either *publish* messages to, or *subscribe* to, in that you receive any messages from a given publisher. *MQTT clients* (like the phone app you'll install to send GPS coordinates) publish their messages to *topics* — different channels on the broker to which a client can publish or subscribe (Figure O).

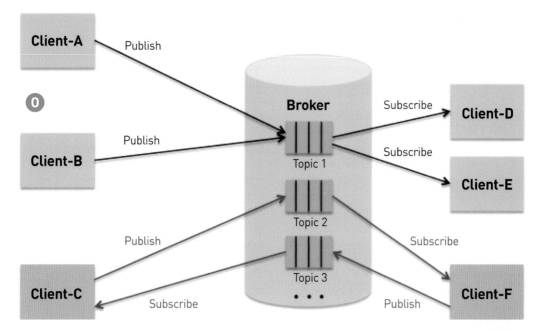

There are several different options for setting up an MQTT broker:

- **Self Hosted:** You set up MQTT broker software on a server you control and operate; Mosquitto is a popular option.
- **Paid Service:** You pay for an MQTT connection where you control the topics and messages you publish/subscribe to; CloudMQTT is one example.
- **Open Broker:** You post to a free and open public MQTT broker such as iot.eclipse.org. Great for testing but not secure!

Each type has its benefits and downsides. Check the project page at makezine.com/go/weasley-clock for a fuller discussion.

After you set up the MQTT broker that works for you, you'll want to create an MQTT user for each person who'll be represented on the location clock, and one for the Raspberry Pi to use as well. MQTT doesn't require authentication for publishing or subscribing out of the box, but it's very easy to set up authenticated topics so that only authenticated users can publish or subscribe to specific topics.

Keep all this user login information handy — you'll use it again when you're configuring OwnTracks and Node-RED to connect to the MQTT broker.

WARNING: Your phone will be publishing your GPS location coordinates and timestamps with a fairly high level of accuracy, so you'll want to make sure you're using secure SSL connections with adequate certificates and authentication!

CONNECTING PHONES TO MQTT BROKER

For GPS services on this project, I decided to use OwnTracks, a free app for iOS and Android devices. It's very easy to set up and start your MQTT connections.

To set up OwnTracks on Android phones or iPhones, follow the instructions I originally published at instructables.com/Build-Your-Own-Weasley-Location-Clock. Here's a quick list of what goes where between MQTT and OwnTracks:

OwnTracks	MQTT
Host	Server
Port	Port (standard, not SSL)
Use Websockets	Keep disabled
Authentication	Enable
Username	User
Password	Password
Device ID	Whatever you want
Security Options	Keep TLS turned off

Once you have the correct settings put in, bring up the Websocket UI on your MQTT console. If everything is set up correctly, you'll get a new MQTT message when your phone connects to your MQTT broker and updates its location.

SETTING UP LOCATIONS IN OWNTRACKS

Now that you've connected OwnTracks and your MQTT broker, you can set up "geo-fence" areas that will be the trigger points for when the clock hands should turn.

On the OwnTracks main page, go to the menu and select the Regions option (Figure **P**). When the region window comes up, hit the plus sign at top right to add a new region. Give the region a name in the Description field (Figure **Q**) and keep track of it — you'll have to reference it when you get to the setup on the Raspberry Pi.

You'll need to also supply your latitude and longitude coordinates. The newer versions of the OwnTracks app for iOS and Android link to Google Maps so that you can input an address and pull the lat and long coordinates from there.

You'll also have to specify the radius of the region (in meters). The region can be as big or small as you want. One of the regions I set up is the entire city of Chicago!

> **IMPORTANT:** At the very bottom of the setup page, make sure you check the option to Share. If you don't have this turned on, then if you enter or leave a region, the region name isn't sent in the MQTT update, so your clock won't know where you are.

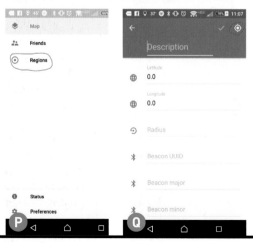

NODE-RED WORKFLOW

The Node-RED flow for this setup is fairly simple (Figure **R**), but took me a good while to figure out. The flow starts with the connection to the MQTT broker with the *MQTT node*. This node is pretty self explanatory, but make sure your MQTT node is subscribed to the correct topic. For my clock, I wanted the hands to move only during OwnTracks transition events (right when you enter or leave one of the waypoints). OwnTracks creates a separate topic for these events in the format *MQTT-topic/event*, so for example, mine was *owntracks/userID/deviceID/event*). As long as you subscribe to that topic with your MQTT node, you should only pick up messages when you enter or leave one of your waypoints.

When the MQTT node is triggered from the broker, a JSON string is sent to Node-RED with all the necessary info from your device. JSON (JavaScript Object Notation) is just a data-interchange format that organizes all your content in a manner that's easy to read both for you and a machine. Here's one of the JSON strings that was sent from my broker:

```
{"_type":"transition","tid":"at","acc":
18.788334,"desc":"Home","event":"enter"
,"lat":00.00000,"lon":00.000000,"tst":1
480793788,"wtst":1477007711,"t":"c"}
```

For my clock, really the only parts of this string I need are the location (**"desc"** for description) and whether I'm entering or leaving (**"event"**). I can't really do a whole lot with just a string, so I need to convert the string into an array of JavaScript Objects.

To convert this string into objects, use the *JSON function node* that comes standard with the *node-red-pi* installation. This parses through the JSON string for you, and formats each string into a group of objects.

For this setup, the JSON string parser feeds directly into a *switch node*. This node takes the newly created object `msg.payload.event` and checks whether that object is set to `leave` or `enter` (Figure **S**). If it's set to `enter`, then the flow is directed to a second switch node that checks the value of the object `msg.payload.desc`; this switch node then feeds into whichever location matches up with the **desc** object value so

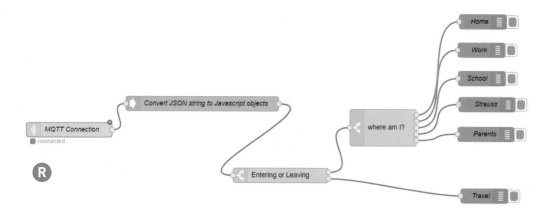

that it can move the hand.

If in the first switch node the `msg.payload.event` object is set to `leave`, then the flow immediately moves to setting the clock hand to "Travel" (Figure **T**). I then have Node-Red run a system callout (like if I was calling something from the terminal) that runs a Python script, and passes a variable of the `msg.payload.desc` location. My Python script picks up the variable and knows where to move the clock hand based on that.

When the flow moves the clock hand to "Travel," a *trigger node* is also activated that counts for 6 hours. If `msg.payload.event` hasn't been updated (from when I enter another known location) after 6 hours, then the flow activates the servo to move the hand to "Lost" and sets another trigger. The second trigger counts for 24 hours while waiting for an updated `msg.payload.event`. If it doesn't receive that update in 24 hours, it finally moves the hand to "Mortal Peril." Bit of a gag if I have to go on a trip for work or if I turn my phone off for a few days, but other than that, should be pretty clear!

OFF TO PLATFORM NINE AND THREE-QUARTERS!

With the clock hands set in place, and the Python code loaded up, my Weasley location clock is all set (Figure **U**). Off we go into the wizarding world, with our magic whereabouts clock keeping tabs on everyone in the family.

I hope you'll give this project a try, and share your improvements at makezine.com/go/weasley-clock. ◐

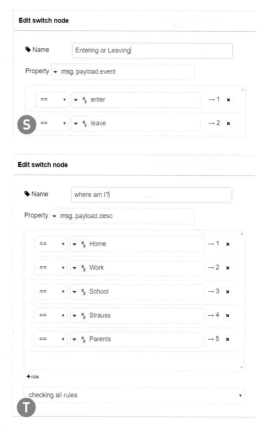

Maker Magician

A card mysteriously rises, and in a flash, changes into the right one!

Written by Mario Marchese

This project is an adaptation of "Rising Card From Envelope" by Bill Severn and Pete Biro, *Tarbell Course in Magic Vol. 7*, page 121.

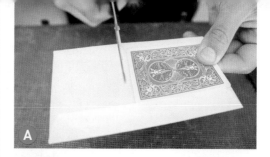

TIME REQUIRED:
15 Minutes

DIFFICULTY:
Beginner

COST:
Under $5

MATERIALS
» **2 packs of the same playing cards**
» **Envelope** I used a 3⅝"×6½" envelope.
» **A bunch of rubber bands**

TOOLS
» **Pencil**
» **Markers or crayons**
» **Scissors**

MARIO "THE MAKER MAGICIAN" MARCHESE is an all-ages theater and virtual performer who has appeared on *Sesame Street*, NBC's Universal Kids, and live on tour with David Blaine who calls him "my favorite kid's magician of all time!!" mariothemagician.com

THE EFFECT

The magician has a spectator select a random playing card. Then they have the spectator place the card back in the deck, *and* have the spectator mix up the deck. The magician places the cards into a contraption called a Card Machine. A card mysteriously rises, but it's the wrong card. Within a flash the wrong card instantly changes into ... the selected card!

THE BUILD

PREP THE ENVELOPE

1. Seal your empty envelope.
2. Line up a playing card along the side of the envelope, for sizing, and cut the envelope about ½" past the card (Figure **A**).
3. Turn the envelope so the open end faces up. About ¾" below the open end, cut or tear small half circles on either side (Figure **B**). This is so a pencil can slide through, but size-wise the hole should be wide enough to fit two pencils.
4. Outline both sides of your envelope with a dark color (Figure **C**).

Now, make it your own! Create the look of your Card Machine! I chose to draw gears on one

Katie Rosa Marchese

side, with images of a 9V battery, wires, buttons, and switches (Figure **D**). You can do something similar or make the design completely your own. On the other side of the envelope I wrote "Card Machine" (Figure **E**). Get creative! Think about how you want your Card Machine to look, and think also of the overall look of your show. Your props should all look like they belong with each other. They don't need to "match" exactly, but they should complement each other: battery and wires, or stars and planets, or birds and flowers, or whatever. This helps create a cohesive theme to your show.

HACK THE PENCIL

1. It's gimmick time! Take a few rubber bands and twist and wrap them onto the middle of your pencil. You'll need to add enough rubber bands to create about ¼" of rubber banding all around (Figure **F**). I have three or four wide rubber bands on my pencil.

2. Slide your pencil carefully through the two holes you created in the envelope, so that the rubber banded area is now hidden inside the envelope (Figure **G**).

THE PERFORMANCE

Now, how do we know which card is selected? How do we perform the trick? The secret is that we need to do something called *forcing a card*. That means that we'll actually *make* the spectator choose the card we want them to choose.

We need two of the same exact card. That's why this trick requires us to have two matching decks of cards! Make sure the pattern on the backs of the cards is exactly the same. The first step is to choose which card you want the audience to pick for the trick. I chose the jack of hearts, for example.

1. Take one jack of hearts from one pack and slide it into the Card Machine, so that the back of the card is against the back wall of the inside of the envelope, and the face of the jack is against the rubber bands (Figure **H**).

2. We don't want the audience to ever see this extra card. When the Card Machine is resting on your table, always make sure the opening of the envelope is facing you. When you pick up the Card Machine for your performance, you'll want to pinch both the jack of hearts and the envelope together (Figure **I**).

3. The second jack of hearts goes on top of the deck of cards that you will be using in the routine (Figure **J**). That will be your *force card*.

So, you have the jack of hearts on top of the deck. You have your Card Machine resting on your performance table with the opening facing you. And the second jack of hearts is hiding in the envelope. You're ready to go!

4. Hold the cards facedown in a *dealer's grip*, as shown in Figure **K**.

Remember, the jack of hearts should be on top, facing down.

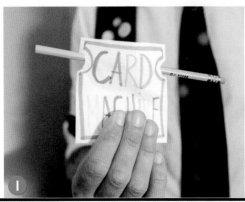

Katie Rosa Marchese

5. Have a spectator pick up a small portion of cards from the top of the deck, flip them over, so they are now faceup, and put them back on top of the deck, still faceup (Figure **L**).

6. Say, "This is a random card!" pointing to the one on top.

And say, "Now, pick up a bigger pile of cards, flip them over, and put them on top again" (Figure **M**).

"This is another random card!"

7. Slowly spread the deck faceup, showing the audience that the cards are all mixed up, some faceup, some facedown (Figure **N**).

8. When you reach the first facedown card, stop, and have your spectator memorize that card and

show the rest of the audience, while you avert your eyes (Figure **O**).

That will be the jack of hearts, and your audience will not suspect that the selected card was not random. You've officially forced a card! Try it out yourself!

NOTE: This method of forcing a card comes from "The Cut Deeper Force" by Ed Balducci and Ken Krenzal; *Hugard's Magic Monthly* Vol. 14, No. 6, page 502; November 1956.

9. After the audience has seen the card and the spectator has memorized it, give someone else the pack of cards, ask them to fix the cards and shuffle the selected card in, making sure it is not

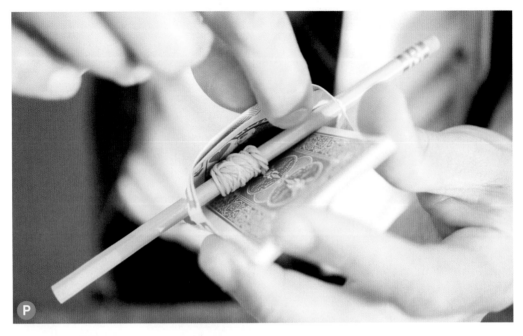

on the top or the bottom of the deck, but mixed somewhere within. When everything is shuffled, take the cards back, and explain that you have no idea what card they picked nor where it is in the deck!

10. Now, pick up the Card Machine, pinching the second jack of hearts and envelope.

11. Take a pack of around 15 cards from the top of the deck, and insert them in the Card Machine, on the opposite side of the pencil from where the jack of hearts is hiding. Make sure the cards face the same way as the jack of hearts (Figure **P**).

12. Turn the envelope around, without showing the inside, and hold the Card Machine by the pencil on each side (Figure **Q**).

13. Ask your spectator to name their card aloud! After they do so, count, "One, two, three," as you move your arms, down to up, with each number. On the third movement, twist the pencil *toward yourself* (Figure **R**). The wrong card will rise!

> **TIP:** Magicians often pretend to fail, so that when the magic finally does happen, the surprise is even greater!

14. Ask, "Is this your card?" Obviously, they will say no.

15. Keep that card in the Card Machine, risen about three-quarters up. Move your arms up and

T

U

V

W

down again, as you count to three again. This time, swiftly twist the pencil *away from you*, and two things will happen simultaneously ... the wrong card will spin down into the envelope, and the correct card will rise up (Figure **S**) — the jack of hearts!

As always, practice this move in front of a mirror. When you coordinate your movements well and twist the pencil with the right amount of speed, it really does look like magic! Experiment with how many cards to put in your Card Machine and how many rubber bands to wrap around your pencil to find the perfect combination to create the level of pressure and grip needed to successfully perform this move.

ALTERNATE VERSION

Essentially, we've created a gear with the rubber bands around the pencil. Hack it, tinker with it, think of new ideas! Create a routine of your own!

Maybe you want the right card to rise right away! If that's the case, place a chunk of cards

right *behind* the jack of hearts. Experiment with this: Instead of twisting the pencil with your fingers, hold onto the pencil with one hand, and flick the envelope with your other fingers, so it spins around the pencil. If you spin the envelope with the right force, and in the right direction, by the time the envelope makes its full revolution, the jack of hearts will be sticking out (Figures **T**, **U**, **V**, and **W**)! Centrifugal force keeps the cards in place upside down. The spinning causes the "gear" to push the selected card out. ◗

Coffee Roaster Wobble Disk

Written and photographed by Larry Cotton

LARRY COTTON has finally given up on doing anything earthshaking. He loves electronics, music and instruments, computers, birds, his dog, and wife — not necessarily in that order.

(A)

Paddles

(B)

Fabricate the eccentrically optimum bean stirring device for a perfect roast

The Simple Sifter Coffee Roaster in *Make:* **Volume 71 does a fine job of roasting coffee.** But it can be made even better by replacing the bean-agitating paddles with a wobble disk (Figure Ⓐ). No other coffee roaster can claim this unique method of stirring the beans. If you haven't yet built one (makezine.com/projects/simple-sifter-coffee-roaster), build it with the wobble disk instead of paddles; you'll be glad you did.

Here's how to replace the paddles with a wobble disk.

1. DISASSEMBLE

Remove the paddles and shaft (shown assembled in Figure Ⓑ) from the Simple Sifter by reversing assembly instructions. You shouldn't have to pull the shaft (axle) all the way out if its straight portion is threaded. The motor can be loosened, but doesn't need to be completely removed.

Most shafts are fully threaded except for the cranking handle end; they are easiest to mount the wobble disk on. Others come with a few threads on the end opposite the handle, for a nut. If you have one like that, you can cut threads (usually with a 10-24 threading die) on the first few inches of the smooth part of the shaft. Clamp the shaft in a vise first, of course.

> **NOTE:** 10-24 seems to be the standard thread for 8-cup flour sifter shafts. If yours differs, copy it. Dies are available in most hardware stores.

If you want to make an entirely new shaft from 10-24 all-thread (threaded rod), just copy the dimensions from the existing shaft. Threads on the motor-driven end of the shaft won't interfere with the roaster's operation, but you can sleeve it if you wish. (Telescoping antennas are an excellent, cheap source of small tubing!)

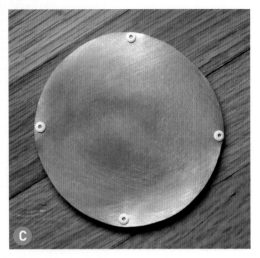

C

disk-to-shaft mounting bracket

1-1/8"
9/32"
7/32"
drill 7/32" thru
1/4"
before bending
1/2"
1/16"
drill 9/64" thru
before bending
9/16"
45°

D

2. MAKE THE WOBBLE DISK

Cut a 5¼" diameter wobble disk from ⅟₃₂"-thick sheet aluminum or steel. Aluminum can be cut with metal-cutting shears, but for steel I used a jigsaw with a metal-cutting blade, firmly clamped upside down (blade pointing up) in a vise. (I don't own a metal-cutting bandsaw, which should be ideal.) In any case, be very careful and wear safety glasses!

An alternative is to use several sheets of thinner (.010") flat metal, such as aluminum flashing, riveted or screwed together (Figure **C**). The finished wobble disk needs to be stiff; bendy ones cannot push the beans around with authority!

Smooth the outside diameter of the disk, working up to about 320-grit sandpaper. Drill an elongated hole in its center (measure carefully) to allow the disk to be mounted on its shaft at a 45° angle. Here's how: Use a ⁷⁄₃₂" bit, preferably in a drill press, to drill the hole. When the bit breaks through, and while it's still turning, hold the disk tightly, and slowly tilt it back and forth to 45° angles. Presto — an elongated hole! Or if you don't feel comfortable with that technique, lengthen the hole (both ways) with a small rat-tail file until the disk tilts at a 45° angle on its shaft.

3. MOUNT DISK TO SHAFT

There are two ways to mount the wobble disk on the shaft: with a small bracket or with two short pieces of aluminum tubing.

E

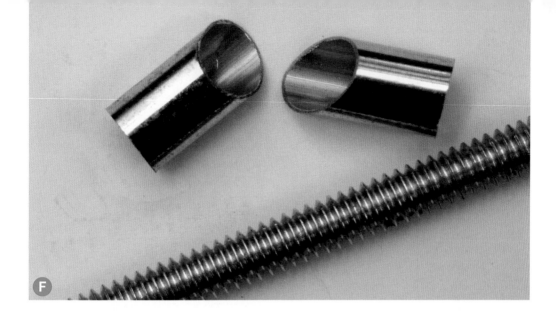

3a. If you have a small piece of ¹⁄₁₆"-thick aluminum, make a disk-mounting bracket to Figure **D**. If you don't have the aluminum, skip to Step 3b.

Push the disk about halfway down the shaft. Thread a 10-24 nut until it touches the disk. Push the bracket over the shaft until its longer (bent) end is flush with the disk. Mark on the disk through the small hole in the bracket, remove the bracket, and drill a ⁹⁄₆₄" hole where you marked it.

Put the parts back onto the shaft in the same order, then fasten the disk and bracket together with a ¼"-long 6-32 Phillips screw and nut (Figure **E**). Go to Step 3c.

3b. For this disk attachment method, use 1" or so of approximately ¼" ID aluminum tubing. Cut with a hacksaw two roughly ½" lengths, each with a 90° and a 45° end. One 45-degree hacksaw cut is all that's necessary (Figure **F**).

Thread one 10-24 nut onto the shaft near the middle. Slip one piece of the aluminum tubing onto the shaft to the nut, with the angled end away from the nut. Slip the disk onto its shaft and seat it against the angled face of the tubing.

Add the second short piece of tubing (angled end first) so the disk is trapped at a 45° angle. It's OK for the tubing pieces to be slightly askew relative to the shaft!

3c. Regardless of the disk-holding method, thread another 10-24 nut onto the shaft so that it traps the disk as close to the center of the shaft as possible. Holding both nuts with wrenches or pliers, tighten them until the disk is firmly mounted.

Page 56 shows the tubing method holding the disk. If your tubing material is slightly too big in diameter, add #8 washers between the nuts and the pieces of tubing.

4. REASSEMBLE AND ROAST!

Reassemble everything, keep fingers clear, run the motor and check (listen!) for any spots where the disk might drag the sifter sieve. No rubbing? You're good to go.

If the disk does touch the sieve occasionally, there are three things you can do:

- Cut a 6"-long, 1"-diameter wood dowel and round one end with coarse sandpaper. Gently press the rounded end into the interfering area(s) of the sieve.
- Reposition the disk, with its mounting hardware, away from the interference.
- Re-cut the disk ⅛" smaller diameter. (Flour sifters' sieves aren't identical, unfortunately.)

Happier roasting! ✪

Build your Simple Sifter Coffee Roaster at makezine.com/projects/simple-sifter-coffee-roaster, and more of Larry's awesome projects at makezine.com/author/larry-cotton.

Written by Tom Whitwell; original project by Bryan Boyer

Very Slow Movie Player

Ambient cinema is the perfect lockdown project

TOM WHITWELL is managing consultant at Fluxx (fluxx. uk.com/books). He also designs electronics as Music Thing Modular (musicthing.co.uk).

BRYAN BOYER is cofounder of Dash Marshall (dashmarshall. com), where he leads the Civic Futures urban innovation consultancy. He's also founding director of the Urban Technology program at University of Michigan's Taubman College of Architecture and Urban Planning.

In December 2018, Bryan Boyer published "Creating a Very Slow Movie Player" on Medium. It's a wonderful essay about light and architecture and Brasília. Boyer describes building an e-paper display that shows films at 24 frames per hour, rather than 24 frames per second. So it would take about a year to play the 142 minutes of *2001: A Space Odyssey*.

I thought about Boyer's essay often. I visited the e-paper department at Waveshare, a Shenzhen-based electronics retailer, and when they offered a 7.5-inch e-paper screen with all the connections for a Raspberry Pi, I bought one and got it working over a couple of days.

It's been playing *Psycho* in the corner of our dining room for months. I set it to run slightly faster than Boyer's — it refreshes every 2 minutes, and jumps forward 4 frames each

TIME REQUIRED 3–4 Hours

DIFFICULTY Intermediate

COST $150–$200

MATERIALS

» **Waveshare e-ink display HAT for Raspberry Pi, 800×480, 7.5"** waveshare.com #13504
» **Raspberry Pi 4 single-board computer, 2GB** with power supply. You can get both in our Getting Started with Raspberry Pi kit, makershed.com/collections/raspberry-pi.
» **microSD card** I bought a pre-installed NOOBS 64GB card, shop.pimoroni.com #RPI-077.
» **Picture frame**

TOOLS

» **Computer with internet connection**

time. That's about 2 minutes of screen time per 24 hours, a little under 3 months for a full 110-minute film.

Psycho is full of visual treats, which reveal themselves very gradually. Some images — like Janet Leigh driving — stuck around for weeks, while the shower scene was over in a day and a half (Figure Ⓐ).

BUILDING YOUR OWN VSMP

If you want to build one for yourself, there are four things to do:

1. Get the Raspberry Pi working in headless mode without a monitor, so you can upload files and run code.

2. Connect to the e-paper screen (Figure Ⓑ) and install the driver code on the Pi.

3. Write some code (or use mine — link below) to extract frames from a movie file, resize and dither those frames, display them on the screen, and keep track of progress through the film.

4. Find some kind of frame to keep it all together.

This is a relatively straightforward project. There's no soldering and no hardcore coding. If you're at all comfortable using a command line, and you've seen a Python script before, then you'll be fine. You can build my version at debugger.medium.com/how-to-build-a-very-slow-movie-player-in-2020-c5745052e4e4 and see Bryan's original at medium.com/s/story/very-slow-movie-player-499f76c48b62. Or build a new version! You can share it at makeprojects.com. ⊘

TIME AND LIGHT

"The screen technology used in VSMP is reflective like a Kindle instead of emissive like a television or computer, which means that the image is always a compromise with the environment. If the room is dark, VSMP's imagery is dark. If the light is warm, VSMP is warm. If the room is bright, VSMP is bright. The device's interactions with shadow and ambient light are its most distinguishing feature."
—*Bryan Boyer*

Bryan Boyer; Tom Whitwell

Stick Man

Show your Lego love by constructing an oversized wooden minifig

Written and photographed by Simon Begg

 SIMON BEGG is a young professional woodturner from Sydney, Australia. He mainly focuses on carved embellishments, German ring turning, and making any strange projects that people are after.

These are some giant Lego men that I have made. I built them as a challenge — I like doing a project like this every year to push myself and learn from the design process. These have all the movements, clips, and operations that a standard Lego minifig has. They are to scale (1:6.25), and I built them using Huon pine, Australian red cedar, and American walnut. Here's how to do it.

DESIGN AND TIMBER SELECTION

Before you start building one for yourself, design and timber selection is very important.

I copied the designs directly from the web (Figures **A** and **B**) and measured actual Lego pieces. It's pretty simple to get the measurements, but there are a few subtle design features that I never realized until I studied these:

- The legs taper outward. Not much, but the distance at the bottom of the feet is equal to the bottom of the body. The top of the legs sit a few millimeters underneath.
- The leg mechanism is not flush at the front or back
- There is a flat spot on the arms
- There is a cutout on the legs that houses the middle disk.

These are all really important, and make a big difference. I even tried slightly changing the sizings on the face and it looked so wrong. They are well designed.

Choosing your scale is also important. I chose 1:6.25 because that gave me the biggest size with the timber I had. The body thickness is 50mm (standard timber sizing). By luck, at this scale the hand also holds a real Lego man perfectly.

Timber selection: The main thing is timber density. With dowel joints that are tight but not too tight, soft timbers are best. Color is only really important for the head. Try to get a yellow timber. The other colors are not as important but something contrasting is good (Figure **C**).

TURNING THE HEAD

The head is simple turning on the lathe (Figure **D** on the following page). Get your diameters and markings right with the calipers. Use the parting tool to get accurate sizings. Roll the radius with a skew or a gouge. After that's done, drill a 30mm hole in the bottom for the connection to the body.

TIME REQUIRED:
~25 Hours

DIFFICULTY:
Moderate

COST:
$30–$40

MATERIALS
You'll need a selection of wood that fits the scaled version of your minifig. I used:

- » **Huon pine** for the head and hands.
  Head: turned to 60mm diameter, 70mm tall
  Hands: turned to 32mm diameter, 27mm tall
- » **Australian red cedar** for the body and arms.
  Body: 100×80×50mm
  Arms: turned to 32mm diameter, 75mm long
- » **American walnut** for the legs and face.
  Legs: 90×45×45mm; Leg plate: 95×48×13mm
  Face: veneer scraps
- » **Dowels**
- » **Glue**

TOOLS
- » **Wood lathe**
- » **Bandsaw**
- » **Linisher sander / belt sander**
- » **Palm sander**
- » **Drill press**
- » **Forstner bits**
- » **Utility knife** aka Stanley knife

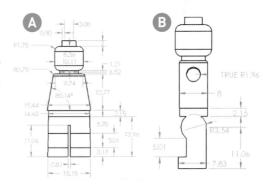

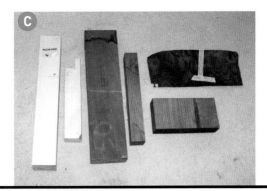

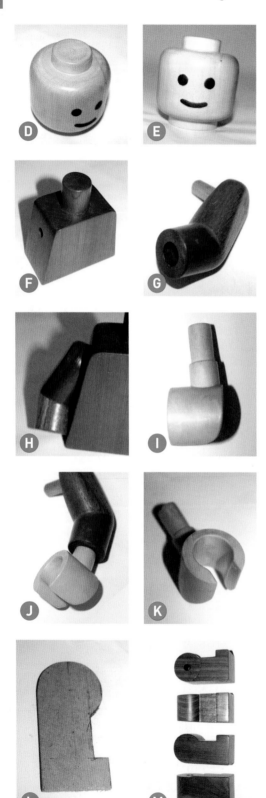

THE FACE

I made my facial features with a walnut veneer (Figure **E**). The eyes were easy, punching them out with a plug cutter. The mouth was harder, having to cut that out by hand with a Stanley knife. It's important to sand the edges. I carefully glued on the pieces with PVA glue, using clear Sellotape (aka Scotch tape) to hold them on. You'd normally use masking tape but this allowed me to ensure the position was exact.

THE BODY

The body is another simple part. The shape is easy to get from the designs. I used a bandsaw to cut it out and then sanded it up. The arm holes are central in width and not too far from the top. Drill the two 30mm holes on the bottom to clip in. The dowel at the top has to be made. This has a 30mm diameter that matches the hole in the head, but it's 29mm at the top. This taper allows the head to slip on with ease. Also drill a 30mm hole in the body to fit this dowel (Figure **F**).

THE ARMS

The arm is two parts but comes from one turning. A cylinder is turned with a taper with a round on top. Partway down the arm, cut it at an angle of 67.5° with a thin blade. Rotate it 180° and reglue. This should create a finished angle of 135° (Figure **G**). Drill holes for the hands and for the dowel to connect to the body (Figure **H**).

THE HANDS

There are two parts to the hand. The wrist dowel is turned with two diameters, one that fits into the arm and the other a little larger to fit into the hand (Figures **I** and **J**). The hand itself is first turned to the correct diameter. Drill a hole in the side that fits the wrist dowel and glue it in. When dry, you can drill out the middle of the hand. Make two cuts in the front of the hand with your chosen saw (Figure **K**). Radius the bottom of the hand using a sander.

THE LEGS

The easiest way to get the legs consistent is to make a template (Figure **L**). The round at the top has to be spot-on, as this moves in the leg mechanism. Cut out the shape on the bandsaw

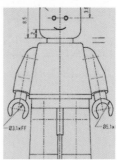

and sand to shape (Figure M). Drill out the centers for the dowel using a hole saw to create the small recess, then increase the hole with the 10mm bit for the dowel. Chisel the rest of the recess flat.

The legs may need adjusting when you make up the mechanism. Sand the taper later, after making the mechanism, so you can make it exact.

LEG MECHANISM

This is the really hard part. First, cut the center block. This can be left a bit oversized, but make sure it's the correct thickness. Next, turn the two body dowels to the same diameter as the one for the head. These are glued into a recess. It's a tight fit so they hold well. Drill two holes at the top for this. These have to be exact. The joint will be really tight and hard to separate but it needs to be tight enough to hold well. Glue these in (Figure N). I used epoxy to be sure.

Next, shape the curve on the underside of the block. To do this I made a sander of the correct diameter by turning pine on the lathe into a cylinder slightly smaller than required diameter, and then gluing abrasive paper to it.

I shaped the disc on the bandsaw and sander. The disc has a tenon on it: drill a hole in the center and glue a dowel halfway though. Then drill and chisel a mortise into the center of the block and glue the disc in (Figure O). This part is so hard because it has to be exact to get the legs to work well (Figure P).

SANDING, FITTING ALL THE PARTS, AND FINISHING

Sanding, sanding, and more sanding. It has to be done. Some bits can be done with a sander but there are also parts that need to be done by hand (Figure Q). Anything that is too tight, sand. Make sure it all moves. It does have to be a bit loose because the lacquer will tighten it up.

For the finish I chose lacquer, for a well-wearing finish that will protect the timber. I sprayed on the lacquer, and I used beeswax on some of the joints to help with movement.

ENJOY!

Once finished, you can play with these! They are a lot of fun and bring back childhood memories. ❂

Corona Creativity

Cooped up by Covid-19, makers are sharing fun and functional fixes that you can make, too

Written by Keith Hammond

A

DISTANCE LEARNING

Ⓐ LIGHTBOARD FOR ZOOM TEACHING
When USC professor Emily Nix tweeted her DIY lightboard — like a see-through chalkboard — Alex Hollingsworth ran with it and shared a how-to on YouTube. Top tips: Edge-light the plexiglass with LEDs to make your writing shine, flip the image in OBS software so it's not backward, and shine a 'key light' on yourself. twitter.com/EmilyNix100/status/1300541067209531392 and youtu.be/YetwRAEVS_U

Ⓑ $3 DOCUMENT CAMERA
For distance learning, kids have to submit photos of their work, and using the laptop camera is just awkward. Kindergarten teacher Andres Thomas's 3D-printed clip-on mirror transforms it into an instant document camera. He's got a cardboard version too. instructables.com/3-Clip-on-Document-Camera

WORK FROM HOME

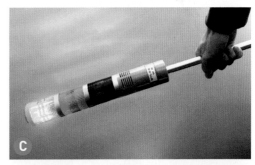

Ⓒ HOME OFFICE STATUS ALERTER
Light the beacon for your family: Green = I'm free, Yellow = I'm busy, and Red = Def don't disturb! This project is a glorified IoT demo by developers at BalenaCloud, but I love that big industrial LED tower, controlled by a Raspberry Pi running a web page you access from desktop or phone. As seen at Virtually Maker Faire 2020. makerfaire.com/maker/entry/71457 and youtu.be/gcj6nA7IAXE

Ⓓ BLUETOOTH ZOOM COMMANDER
Need to mute or kill the camera quick? George White built this little Bluetooth Zoom keyboard "based on an Adafruit NeoTrellis and seeded with code from John Edgar Park. Still more work to do, but having the main Zoom features as single presses is happy-making." github.com/stonehippo/zoom-keyboard and twitter.com/Stonehippo/status/1328423629781131264

```
cord.value and not pulled
nd(Keycode.COMMAND, Keycc
t("ENDING ZOOM CALL!!|")
edSwitch = True
.sleep(0.4)
```

E ZOOMOUT "END VIDEO CALL" PULL CORD

"I always awkwardly struggle to get to the 'end call' button on video calls," says Brian Moore. So he used an Adafruit Feather Bluetooth board to make his Zoomout pull-chain switch. "It sends a key command to my computer, and then Alfred app runs an AppleScript based on that hotkey that kills any open Google Meet tabs or Zoom calls." twitter.com/lanewinfield/status/1339257875034566656 and github.com/lanewinfield/zoomout

OUT AND ABOUT

F GLAMOUR FACE MASK

If you gotta wear it, then wear it well. Sewing author, YouTuber, and pattern designer Gretchen "Gertie" Hirsch shared this vintage burlesque-inspired mask in two versions, one with Chantilly lace and one with French veiling and rhinestones. Sew yours at youtu.be/Rgz_rPeSn94 and charmpatterns.com/shop/glamour-mask-free.

G WEARABLE PROJECTION

Thai artist Natthakit "Kimbab" Kangsadansenanon was already tinkering with a 3D-printed personal projection rig, using the surrounding environment as a canvas. Then social distancing gave him his first topic: a "private zone" cast upon the sidewalk

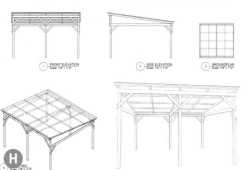

around him. Also good for projecting nighttime traffic cautions for pedestrians or cyclists, or "I am single" if the bars ever reopen.
makeprojects.com/project/wearable-projection

GET TOGETHER

⊕ PARKLETS FOR OUTDOOR DINING
Indoor dining is outlawed in many places, but not every eatery has room for outdoor seating. "Parklets" to the rescue! Art and events producer Paul Belger has pivoted from building Burning Man spectaculars to helping restaurants build these little outdoor cafés in parking spaces on city streets. He's shared this basic plan at makezine. com/go/parklet.

⊕ HUGGING BOOTHS FOR SENIORS
One of the saddest sights of the pandemic: elderly folks quarantined behind glass, unable to hold or touch their loved ones. Plastic "hug curtains" appeared last spring and quickly evolved to fully enclosed "hugging booths" — like a big DIY biosafety glove box — so that families can hold each other again. This one by Steven and Amber Crenshaw is my favorite, two-way and simple to build because it mounts in an existing doorway.
makezine.com/go/hugging-booth ⊘

Brian Moore, Courey Ayres/Shameless Pinup, Kimbab, Paul Belger, Amber Crenshaw

Share your Covid creativity with the community at Make: Projects, makeprojects.com.

New From Make: Projects

Makers everywhere are sharing great builds on makeprojects.com — post yours, too!

ANIMOJI MASK
Lenny Leiter
@lenny3000

Through the power of phone app filters and animojis, along with some mirror and projector trickery, this helmet becomes an unforgettable mask for Halloween or any party. The future of faces is already here!

makeprojects.com/ project/animoji-mask

❶ THE INTERACTIVE STORYTELLING RADIO

8 Bits and a Byte @8bitsandabyte

No longer do you have to keep track of multiple pages in a choose-your-own-adventure book. This voice-enabled interactive storyteller is built using a Cold War-vintage Telefunken radio, a Raspberry Pi, and dialogue planning software that follows a decision tree. All you need to do is decide what story to tell and path to take. makeprojects.com/project/the-interactive-storytelling-radio-povqe

❷ BEERCASTER

Jernej Ježovnik @Jerch

It's easy to tune in with and keep tabs on the Beercaster. With some help from some friends, this Telecaster-style guitar was built with over 5,000 can tabs and epoxy. makeprojects.com/project/the-beercaster

❸ TRAINTRACKR — LIVE NEW YORK CITY SUBWAY MAP

Richard Hawthorn @richardhawthorn

You won't ever have to wonder where the F train is again. Through 599 color LEDs, you can track New York subway trains as they arrive and depart from the stations thanks to the Traintrackr custom PCB board. It is built using a HT16K33 LED driver chip and an ESP8266 Wi-Fi microcontroller to update the location data every 10 seconds. Boston and London available too. makeprojects.com/project/traintrackr---live-new-york-city-subway-map

❹ CAT WHISKER SENSORY EXTENSION WEARABLE

Chris Hill @Christian

Built by a Ph.D. student as an extension of a colleague's work, these whiskers translate input from flexible sensors connected to an Arduino Uno to four vibration motors that surround the user's head. It creates a purr-fectly unique experience whenever you wear them. makeprojects.com/project/cat-whisker-sensory-extension-wearable

❺ CASSETTE CYCLE: THE ART OF RECYCLING CASSETTE TAPES & FORGOTTEN TECH

Nicole Majewski @theerrantstitch

Still have some of your treasured mixtapes on cassette or movies on VHS? Save them to build furniture to display your projects, create purses, or to pop into a retrofitted boombox stereo shoulder bag. And don't worry about rewinding before starting your own builds. makeprojects.com/project/cassette-cycle-the-art-of-recycling-cassette-tapes-forgotten-tech

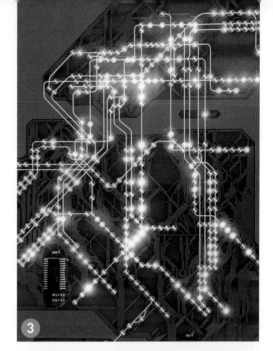

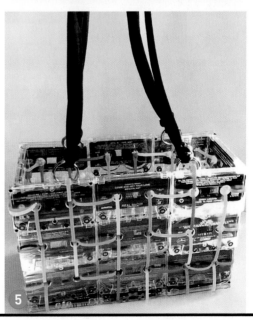

Music is so much more than just rhythm and melody — it's an art with an attraction as universal as it is individual.

Adobe Stock - Iana_samcorp and dlyastokiv, Robin Hails, Liz Clark, Michael Krzyzaniak, Johan von Konow

MUSIC MAKING

AURAL INNOVATIONS: 74

GLOCKEN' ROLL: 84

DR. SQUIGGLES AI RHYTHM BOT: 88

PRINT-A-SYNTH: 98

Music is so much more than just rhythm and melody — it's an art with an attraction as universal as it is individual. Its enjoyment comes from listening to the works of masters, as well as creating our own masterpieces, even those made from combining the most basic of sounds. Musical instruments today are more advanced than ever, but fear not, our maker toolboxes contain the parts to explore their methods and techniques. In this section, we'll first look at how to create your own musical devices using the latest microcontrollers. We'll jump from there into setting up a MIDI-enabled xylophone that you program or play wirelessly with BLE. After that, we'll show you how to make a tapping bot that listens to the rhythms around it and synchronizes its output to match them. And finally, we'll build a pocket synthesizer that omits a dedicated PCB for simple route-in-place wiring. Everything you need to get the band back together, or become a one-person robot band yourself. ❷

AURAL INNOVATIONS

ALL INSTRUMENTS ARE INVENTIONS, AND ALL MUSIC IS MADE UP — SO MAKE YOUR OWN, USING MICROCONTROLLERS Written by Helen Leigh

Humans have been experimenting with sound, making music, and inventing instruments since prehistoric times. In the beginning, we made noises using objects found in nature, including shells and vegetation, but as our skills and tools advanced, so too did the instruments we made.

Every instrument we think about as mainstream today was once a new invention, influenced by a myriad of other instruments from cultures across the world. Let's take the violin as an example, which didn't exist until the 1500s when lute and lyre makers started to tinker with the design of previous stringed instruments. The design of the first violins evolved over the following centuries as the demands of the orchestra changed and new tools and techniques became available.

It's not just instruments either. You might think of classical music mainstays like Stravinsky, Strauss, or Debussy as boring establishment composers, but in their day they were seen as avant garde and radical. The premiere of Stravinksy's *The Rite of Spring* nearly caused a riot when the audience became outraged by the dissonant chords, pulsing rhythm, and modern ballet dancers.

Even the musical notes we use in Western music today were not carved in stone in ancient times: in fact it was only in 1955 that middle A was officially set at 440Hz. Before 440Hz was accepted as the standard for middle A, there were regional and national variations for instrument tuning. Not everyone was enthusiastic about the globalisation of musical notes, and people are still unhappy about the decision to this day! If you spend a little time digging around on the internet you'll find 440Hz conspiracy theories galore, as well as campaigns for middle A to be changed to 432Hz, 438Hz, or even 538Hz.

All music is made up and all instruments are invented, so as makers we should feel empowered to follow in the footsteps of the musical experimenters who came before us and make noises in any way that we want. Who knows where your tinkering will take you?

The magnetic tape recorder kicked off the electronic music revolution in the 1940s.

Robyn Hails, R.G. Jonkman / Wikimedia Commons

MUSIC TECH HACKERS CHANGING HISTORY

Let's take a closer look at one piece of music technology — the tape recorder — and how the people who hacked it changed modern music production. The magnetic tape recorder was invented in Germany and used in World War Two by the Nazis for propaganda purposes. After the war ended, Allied troops found the technology and took it home to their own countries. Eventually, magnetic tape recorders came into use in broadcasting and recording studios across the world. As the technology evolved and became widespread, it also became cheaper and more accessible to people as well as institutions.

As more people got access to these machines, some began playing with them in unexpected ways. In Paris, a handful of these magnetic tape recorder experimenters started sharing their work, leading to the creation of an iconic movement called *musique concrète*.

To edit a recording on magnetic tape, you needed to physically cut the tape with a blade and use sticky tape to attach it to the next part of magnetic tape. The musique concrète folks

HELEN LEIGH is a hardware hacker who specializes in music technologies and craft-based electronics. Say hello to her on Twitter @helenleigh.

Composer and musician Daphne Oram pioneered new electronic music sounds from synthesizers and tape loops in the 1950s, at the legendary BBC Radiophonic Workshop.

would go out into the world and record interesting sounds onto their tape. They would then take sections of this tape, cut it up, flip them (to play the sounds backwards), speed them up or slow them down (for higher or lower tones), distort them, stick them back together, and re-record them to create new exciting new sounds.

Let's take the example of the plucking of a violin string: the sound starts immediately then dies off gradually. In music production that would be called a sharp *attack* and a long *release* or *decay*. Using the new technologies and techniques from the musique concrète folks, you could flip that violin sound around to produce a new sound with a long attack and a short decay. This all seems quite normal to us now, but back then it was revolutionary.

You can actually head over to YouTube now and listen to some experimental musique concrète from the 1940s, but don't expect harmonious compositions or banging beats! Their music may not have made it to the mainstream but their techniques certainly did. Their place in music tech hacking history was cemented in the 1960s, when the Beatles became the first big band to use these techniques to create the otherworldly, psychedelic sounds of their classic track "Tomorrow Never Knows."

> ## "BECAUSE WE WEREN'T EXPERTS WE DIDN'T KNOW WHAT WE SHOULDN'T BE CAPABLE OF DOING."
> **—DESMOND BRISCOE, CO-FOUNDER, BBC RADIOPHONIC WORKSHOP**

THE BBC RADIOPHONIC WORKSHOP

Another lasting effect the musique concrète folks had on modern music was through the work of a remarkable British woman called Daphne Oram. She was a musician and physicist who became one of the iconic figures of early electronic music. She had a job at the BBC as a music balancer, which involved setting up recording equipment and lining up vinyl records for seamless broadcasting. Oram was also a big fan of avant garde music, including musique concrète. She

Musician, hacker, and inventor Sam Battle, aka Look Mum No Computer, gained fame for the Furby Organ — 45 singing toy robots triggered by a keyboard and 90 microcontrollers. He's a prolific builder and sharer of DIY synths and circuits, on YouTube and at lookmumnocomputer.com.

Look Mum No Computer

realised that combining magnetic tape recorders and the new techniques from Paris held enormous promise for new sounds and music. She recorded tones from sine wave generators onto tape loops and experimented with the effects of playing them back at various speeds, creating some of the earliest purely electronic music.

Oram's colleagues at the BBC were not enthusiastic about her work: it took her eight years of experimentation and a lot of persuasion before she could convince somebody to let her create some music for a BBC show. This wasn't just due to the forward-thinking nature of her work, but also because of her gender. "They wanted my work," Oram said, "but they didn't want me." After she teamed up with her colleague Desmond Briscoe, the BBC finally relented to her requests for the creation of a department dedicated to electronic sound. This department was called the BBC Radiophonic Workshop, which became one of the leading experimental sound design studios in the world.

There are a number of documentaries about the BBC Radiophonic Workshop and its alumni that you can watch to explore this fascinating period of music tech history further, including *Sisters with Transistors* and *The Delian Mode*.

INVENTING INSTRUMENTS WITH MODERN MICROCONTROLLERS

The tape recorder is an excellent example of how the wider availability of a new technology and the sharing of experimental techniques led to a string of exciting innovations. We're currently in an extremely exciting time for new, accessible, affordable technologies. We can play with all sorts of microcontrollers and different sensors, and we can use our sensor data to wirelessly control any number of soft synths or digital audio workstations.

It's not just the hardware that is increasingly accessible, but also the knowledge required to use it. Back in Oram's day you would have to travel to a conference or subscribe to a very niche set of journals to find information about emerging music tech. These days we have people like Look Mum No Computer showing us how to build synths on YouTube (such as his insane Furby Organ, pictured above), makerspaces with music hacking groups like London Hackspace's Hackoustic, and educational workshops at events like CCC's Congress.

There's never been a better time to start making music and experimenting with sound!

MODERN MUSIC HACKERS

A few favorites from the Hackoustic group at London Hackspace, hackoustic.org.

① BRENDAN O'CONNOR'S SOUND STITCHER

Sound Stitcher is an interactive vintage sewing machine. Turning the wheel generates digitally amplified sewing machine noises, while interacting with its levers, sliders, and clips causes the sounds to distort and degrade. facebook.com/brendanoconnormusic

② ADRIAN HOLDER'S OBJECT PROJECT

Holder, aka Precis, hacked a turntable so that it not only plays sound the traditional way, with a needle, but also produces music and tones by striking its surface. The turntable also reads the shape, size, color, and movement of objects placed upon it!

③ TOM FOX'S SPRINGYTHING

Springything is an experimental instrument made out of springs, magnets, a coil of wire, and an amplifier. The range and variety of noises and sonic textures that come out of this instrument are quite impressive! Build your own: vulpestruments.com/2017/07/30/springything-tutorial

④ JEN HAUGAN'S DOPPLER MACHINE

The Doppler Machine uses the Doppler effect as a playful way to create new sounds. Haugan plays her instrument in combination with feedback and external microphones, leading to

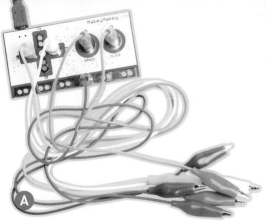

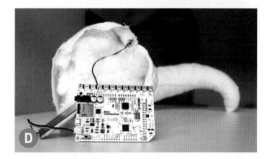

GREAT MICROCONTROLLERS FOR MUSICAL INSTRUMENTS

Makers have made strange and beautiful music with just about every microcontroller out there, but here are some of our favorite devices for inventing instruments.

MAKEY MAKEY

Everyone loves the easygoing charm of the Makey Makey (Figure **A**). It takes seconds to set up with no code or soldering required, after which it transforms ordinary items into touchpads that can be used for whatever fun invention you can dream up, from the infamous banana piano to unusual game controllers. In a Makey Makey circuit, you connect a conductive object (such as a metal item or a thing with a high water content — like fruit or people) to the board with alligator clips then close the circuit by connecting to ground. The Makey Makey then connects to your computer to control whatever you tell it to control.

One of the best ways to use the Makey Makey is as an instrument. I've used the board to play piano with bananas (Figure **B**), play the bongos with jelly (Figure **C**), and trigger samples by taking a sip of my tea. Perfect for beginners or family fun.

BARE CONDUCTIVE TOUCH BOARD

If you have a project in mind that combines touch and sound you should consider giving the Bare Conductive Touch Board a try (Figure **D**). Connect the electrodes on the board to a conductive material (such as copper tape, wire, or conductive paint) and your touch will trigger one of 12 audio samples on the onboard SD card, which are very easy to switch out to samples of your choice.

These attractive, Arduino-compatible boards are reliable and simple. You don't need to look at any code if you just want the basic touch and play functionality, and you don't need to connect it to a computer for it to play your sounds. If you do feel comfortable writing a bit of code, you can use the Bare Conductive Touch Board to sense proximity too — great for all sorts of musical applications! I recently used this board to make a big plush pet tentacle (Figure **E**) that purrs when I stroke it.

Robyn Hails

TEENSY

If you're happy writing Arduino code then some of the best microcontrollers you can choose for music projects are Teensys (Figure **F**). They're tiny, powerful, and eye-wateringly fast, with some excellent advanced audio tools and libraries in the Teensy ecosystem. You'll find Teensys at the heart of many sequencers and synthesizers.

Maker, researcher, and music enthusiast Oscar Oomens paired a Teensy with a Raspberry Pi to create the beautiful SENSEI synthesizer (Figure **G**). The synth features a custom touch interface and wearable gyroscope that allow the user to "shape" sound in multiple dimensions. As well as moving the joystick, you can control the sound characteristics with three force-sensitive resistors on the grip (Figure **H**).

BELA

One of the best ways to make a more advanced musical instrument with embedded technology is Bela (Figure **I**). Bela isn't a microcontroller, it's an add-on for the BeagleBone single-board computer (SBC) [full disclosure: my husband, Drew Fustini, is a volunteer board member of the BeagleBoard.org Foundation]. Bela uses open source software and hardware specifically designed for making beautiful sounding instruments. If you're serious about making instruments, it's a great choice.

I paired a Bela with a Trill capacitive touch sensor to make a series of musical metal circuit sculpture creatures, including a sub bass (Figure **J**) and a harp (page 74).

To learn more about it, let's use Bela to build a rubber duckie synthesizer!

Oscar Oomens, Robyn Hails

SYNTHESIZING SOUND:
MAKE A DIGITAL RUBBER DUCKIE

CREATING INTERACTIVE DUCKIE SOUNDS

Created in the Augmented Instruments Lab at London's Queen Mary University, Bela and the Bela Mini are based on the open source BeagleBone Black and PocketBeagle SBCs. Most SBCs and microcontrollers have add-ons: Raspberry Pi has hats, Arduino has shields, and BeagleBone has capes — such as Bela. It leverages the high-speed capabilities of the BeagleBone to make super responsive, low-latency instruments.

What makes Bela really exciting for me is how it works with not just C++ but also a number of open source sound synthesis programs, including Pure Data and Super Collider. You can use these powerful audio programs to create your noises, then embed them into an instrument that you can play forevermore with no connected computer. Pure Data is a great way to do all sorts of complex and beautiful things with sound.

In this project we're going to use Bela and Pure Data to run a synthesised rubber duckie based

on a *physical model*. With physical modeling, the physics of a specific type of sonic interaction are re-created in code, which lets us synthesize sounds much more realistically. The code in this example uses a physical model of a rubber duck toy, created by Christian Heinrichs. Creating a physical model involves analyzing an object and attempting to understand and then synthesize all the different ways it makes sound — all its different attack and release and intensity and other sound qualities, under a variety of playing conditions such as squeezing the air out fast or slow, letting air flow back in, squeezing too hard, and so on. It's the difference between a simple duckie sample, and a realistically interacting digital duckie that can create its full range of sounds depending on how you play it.

This physical model can be controlled with sensors in real time, making it an extremely exciting way to make rich, satisfying sounds, especially when paired with audio software like Pure Data.

1. WIRE THE FSR

Start by wiring your force sensitive resistor (FSR). FSRs are sensors that detect physical pressure, such as weight or squeezing. We'll manipulate this FSR to engage the duckie sounds (the real rubber duckie shown above is just illustrative).

Insert the two metal pins at the base of your FSR into two separate rows of your breadboard. Next, take your 10kΩ resistor and insert one of its legs into the same row as one of the legs of your FSR. Insert the other leg of your 10kΩ resistor into the ground rail of your breadboard (Figure Ⓚ on the following page).

2. CONNECT THE BELA

Next, wire your FSR to your Bela, which should not yet be connected to power or your computer. Use jumper wires to connect the ground rail of your breadboard to a ground pin on your Bela, then connect the power rail of your breadboard to the Bela 3.3V pin. Find the leg of the FSR that you connected to the ground rail in Step 1. Plug in one end of a jumper wire to a free hole in the same row as your grounded FSR, then connect the other end of the wire to pin A0 on your Bela (Figure Ⓛ on the following page).

3. CONNECT COMPUTER AND SPEAKER

Plug the aux adapter cable that came with your Bela to the audio OUT pins, then use an ordinary ⅛" (3.5mm) aux cable to connect a speaker. Turn the audio level down on your speaker to start with, as the tone we're going to use to test our setup can be loud!

Before you connect your Bela to your computer, take a moment to check that your board is not sitting on anything conductive. The board has metal pins on the bottom, so just make sure you aren't in danger of shorting your board by putting it on a metal laptop, a strip of copper tape, or other conductive material.

Connect the Bela to your computer using the micro-USB to USB-A cable that comes with the board (Figure **M**).

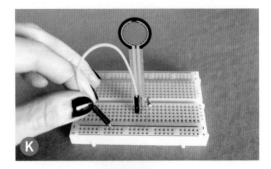

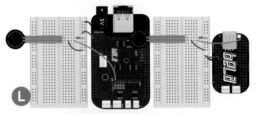

4. TEST YOUR BELA SETUP

One of the nice things about Bela is that you access the IDE (the development environment where you program your board) using your web browser. This means that once your Bela is connected to your computer, all you need to do is open a browser and get coding — no software downloads needed! You don't even need to be online because your computer recognizes your Bela as a local network. After your Bela boots up, go to http://bela.local in a web browser to load the Bela IDE. If http://bela.local doesn't bring up the IDE, try the IP addresses 192.168.6.2 or 192.168.7.2.

There are four main parts of the IDE. The *editor* is where you write your code: C++, Csound, Supercollider Programming Language (sclang), JavaScript, or Pure Data. You stop and run your code using the *toolbar* at the bottom, where you can also find the oscilloscope and the GUI visualiser. The *console* gives you feedback about your Bela, such as sensor data and error messages. Finally, *tabs* is where you manage projects and settings, as well as where you find code examples, an interactive pin out diagram and all the libraries.

Test your setup by navigating to the tabs section of the IDE and running the audio world's version of "Blink": the sine tone. Click on the lightbulb icon in the tabs section to bring up the Examples folder. Click on Fundamentals then

Robyn Hails

click on the example project called *sinetone* to open it up in your editor (Figure **N**). Click on the "build & run" circular arrow icon in your toolbar; your code should start running on the board and if all is well, you'll hear a sine tone.

Sound model of a rubber duckie toy

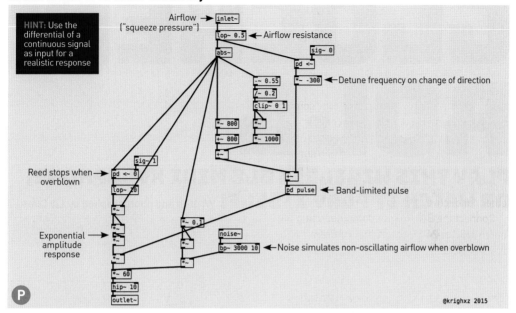

HINT: Use the differential of a continuous signal as input for a realistic response

Airflow ("squeeze pressure") → inlet~

top~ 0.5 ← Airflow resistance

abs~

sig~ 0

pd <~

*~ 0.55 *~ -300 ← Detune frequency on change of direction

/~ 0.2

clip~ 0 1

*~ 800 *~

+~ 800 *~ 1000

+~

Reed stops when overblown → sig~ 1 / pd <~ 0

top~ 10

+~

Exponential amplitude response → *~ 0.1

*~ noise~

+~ bp~ 3000 10 ← Noise simulates non-oscillating airflow when overblown

*~

+~ ← Band-limited pulse / pd pulse

*~ 60

hip~ 10

outlet~

@krighxz 2015

---

5. PROGRAM YOUR BELA DUCKIE

Go back to the Examples tab in the Bela IDE, and scroll down until you find the Pure Data section. Expand that section by clicking on the +, then open the *rubber duckie* sketch. The minimal code you see in the editor area is the main file of the collection of files that make up the Pure Data patch. Unlike the C++ code found in the sinetone example, you can't edit Pure Data patches directly in the browser. We'll take a closer look at the Pure Data code in the next step, but for now you can go ahead and run the rubber duckie code by clicking on the build & run icon in the toolbar.

Press, squeeze, flick, and grip your FSR to try out your new rubber duckie. Try pressing slowly, quickly, firmly, and lightly to see the difference it makes!

6. EXPLORE AND EXPERIMENT

If you want to understand more about the way these difficult-to-synthesize interactive sounds are created, you can start by taking a look around the Pure Data file. Pure Data (often called Pd) is a free and open source visual programming environment for sound, visuals, and other media. It's a wonderful tool for creating complex sound synthesis. You can't edit Pd files directly in the browser, so if you want to a look at the (really

quite interesting!) example code in more detail you'll need to open it with Pd.

First, click on the folder icon in the tabs section of the Bela IDE to open the Project Explorer then download your project and unzip the folder. Next, download and install Pd on your computer from puredata.info. There are two versions of Pd available: Pd Vanilla or Purr Data. Either will work just fine, but I tend to use Pd Vanilla.

Go to the unzipped folder you downloaded from the Bela IDE and open _main.pd in Pure Data (Figure ⊙). Enter Edit mode, then click on the *duckie* and *sigdelta* nodes to take a look around. *Sigdelta* works out the velocity of the sensor stream, making our duckie respond to our touch in a more natural, responsive manner. *Duckie* is the physical model (Figure Ⓟ), which calculates airflow and pressure, as well as detuning the frequency of the duck when you start to release your grip and the "air" starts flowing back in! ⊘

All of my projects in this article — the purring tentacle, the jelly bongos, the banana piano, and the circuit sculpture creatures — will be published in my upcoming book on DIY Music Tech, due out in early 2021. Follow me on Twitter @helenleigh or Instagram @helenleigh_ makes to find out when it's released!

GLOCKEN' ROLL

PLAY THIS WIRELESS BLE MIDI XYLOPHONE, OR WATCH IT PLAY ITSELF!
Written and photographed by Liz Clark

LIZ CLARK, aka Blitz City DIY (youtube.com/blitzcitydiy), is a Massachusetts-based maker who dabbles in electronics, music tech, 3D printing, CircuitPython and anything else that looks interesting that day. When her soldering iron is cooling you can find her with her cats Winnie and Harriet.

My BLE MIDI Robot Xylophone is, as its name implies, a robot instrument. It uses 30 mini solenoids to strike the xylophone (technically, it's a glockenspiel)'s keys. The solenoids are triggered by MIDI data that is received over Bluetooth Low Energy (BLE). The code is written in CircuitPython, which has libraries for BLE MIDI for use with Adafruit's nRF52840 boards (this project utilizes the ItsyBitsy nRF52840).

The main purpose of this robotic xylophone is to play music with the instrument that would not be possible with the standard human method. Mallet instruments can be limited in the music that can be played on them; traditionally, a player uses two mallets to play with a maximum of two-note polyphony, or two notes at a time. More advanced players can use 4 mallets, but this is still limited to only four notes at a time. My robotic modification allows for music to be played on the xylophone that goes well beyond that. Technically, all 30 notes could be played at the same time, although it might not sound very good.

In this project you'll see how you can build complex robotic instruments that are controlled wirelessly with BLE MIDI. If you'd like to build your own, a fully detailed, step-by-step build guide is available at learn.adafruit.com/wireless-ble-midi-robot-xylophone.

WHAT IS BLE MIDI?

BLE MIDI is a wireless protocol that allows an instrument to receive command data. It's great for a large build, like this xylophone, since you can set up the instrument where it fits comfortably, rather than having to be close enough to your computer to be physically connected for data or power. This gives you a lot more freedom, especially for setting up an ideal recording situation with microphones.

BLE MIDI also requires less hardware. Previously to use MIDI you would possibly need an audio interface with MIDI ports. Now, however, most computers have BLE built in, or you can get a cheap BLE USB dongle that's plug-and-play. Once you set up your BLE MIDI device, your software will recognize it as a MIDI device just like a standard USB or 5-DIN connected device.

BUILD THE CIRCUIT

Despite the advanced BLE communication involved

TIME REQUIRED:
4 Hours

DIFFICULTY:
Moderate

COST:
Approximately $200

MATERIALS
- » Adafruit ItsyBitsy nRF52840 Express BLE
- » Solenoids, mini push-pull, 5V (30)
- » Darlington array ICs, ULN2803, 8 channel (4)
- » Port expander ICs, MCP23017, I²C, 16 I/Os (2)
- » Toggle switch, illuminated with cover
- » Breadboard-style PCBs, ½ Perma-Proto (3)
- » Aluminum extrusions, 20×20, 610mm long (2)
- » 3D-printed parts: legs, solenoid mounts, etc.
- » Ludwig Educational Bell Kit or comparable small mallet instrument
- » JST cables, hookup wire, DIP sockets, hardware

TOOLS
- » Soldering iron and solder
- » 3D printer
- » Wire cutters

Adobe Stock - dlyastokiv

in this project, most of the components for the circuit are fairly old-school in the form of two ICs: the MCP23017 and the ULN2803. The MCP23017 is a multiplexer that allows for 16 inputs or outputs to be communicated over I²C. This is how the build can use the ItsyBitsy form factor board but still control 30 solenoids (Figure **A**).

The ULN2803 is a Darlington motor driver. It takes in the logic signal from the ItsyBitsy and outputs the voltage needed to trigger the solenoid. In theory, this means you could have a 3.3V logic level from your microcontroller but still trigger a high voltage motor.

For the circuit, the ItsyBitsy nRF52840 acts as

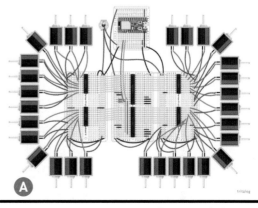

A

the central command. The project is powered via the USB input since the mini solenoids only require 5V. The ItsyBitsy sends the data signals to the two multiplexers over I²C. The multiplexers then output that on/off signal to the Darlington motor drivers, which turn the solenoids on and off to strike the keys on the xylophone.

To keep the wiring nice and neat, I recommend using PermaProto boards and IC sockets so that you can easily exchange the multiplexer or driver ICs if needed (Figure **B**).

THE CODE

I wrote the code for the xylophone in CircuitPython. Each output on the multiplexers corresponds with a specific MIDI note number, matching the physical notes on the xylophone. When it receives a matching MIDI note in a NoteOn message, the solenoid triggers, and after a very short delay it retracts. The code allows for *polyphony*, or multiple notes to play at the same time.

You can download the code on GitHub and in the full guide on the Adafruit Learn system page.

ASSEMBLY

The robot xylophone consists of two pieces of 20×20 aluminum extrusion and 3D-printed parts (Figure **C**), which Noe Ruiz of Adafruit designed.

Two 3D-printed legs hold the two pieces of 20×20 extrusion at an angle so that they run across the middle of each key on the xylophone. This puts the solenoid mounts in perfect position to get the best tone from the solenoids striking the keys. The solenoid mounts are designed to be adjustable so you can raise or lower the solenoids as needed (Figure **D**).

The PermaProto boards have snap-fit cases that you can screw into the 20×20 aluminum extrusions using 3D-printable mounting pieces (or you can make them with aluminum). This keeps the wiring centralized so that the solenoids can easily plug into the main circuit.

The ItsyBitsy and its power connections sit in a 3D-printed enclosure that mounts to the top of the 20×20 aluminum extrusion. On the top of the enclosure, I added a large toggle switch that controls the power flow to the Darlington driver ICs for the solenoids (Figure **E**).

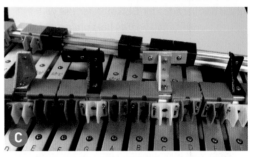

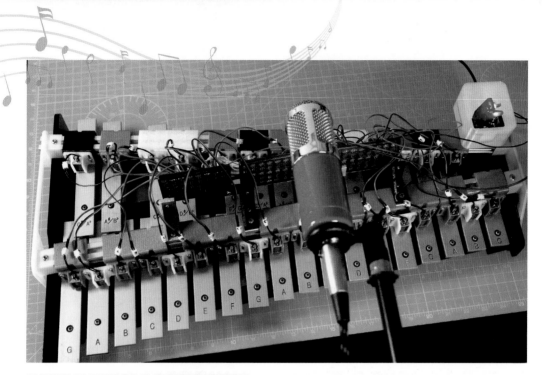

MAKING THE BLE CONNECTION

How you approach setting up your BLE MIDI connection will vary depending on your computer's operating system. If you're using macOS, BLE MIDI is supported natively and Apple has documentation showing how to connect your BLE MIDI peripherals with no additional software.

Windows requires a few more steps. First, to have your digital audio workstation (DAW) recognize your BLE MIDI device as a hardware connection, you need to set up a virtual software connection to essentially "trick" the software. To do this, you can use loopMIDI, which is a software project by Tobias Erichsen.

After this MIDI connection shell is created, you need a software bridge to connect your BLE MIDI device to your DAW or computer-connected MIDI controller. There are a few options to achieve this, but the one tested with this project is MIDIberry. It's a freeware option available through the Windows store that works very well. You can select your MIDI input and output devices in the GUI and it also has a MIDI message debugging stream.

In MIDIberry you can set up loopMIDI as the input and your CircuitPython BLE MIDI device as the output (Figure **F**). This allows the loopMIDI connection to receive the MIDI data inside your DAW and then send it out to your BLE MIDI device.

After the BLE MIDI connection has been established, you can open your preferred DAW and select your loopMIDI connection as the MIDI output.

After this you can play the xylophone live from a MIDI keyboard, or send entire songs to the xylophone and it will play them like a player piano. It's great for recording, and it lets you play pieces that would be impossible when limited to two or four mallets.

CONCLUSION

Adding BLE MIDI functionality to a robotic instrument is a great twist on a classic project. You could use this technique for other instruments as well, such as drums, guitars, or really any instrument that utilizes a hitting or plucking motion to be played. There is a MIDI solenoid drum project that I worked on in collaboration with the Ruiz brothers for Adafruit, also available on the Adafruit Learn system. (Also check out Dr. Squiggles, a solenoid rhythm bot, on the next page.)

Beyond a purely musical application, you could use this technique to trigger motors or other components over BLE. Perhaps you want to turn on a light every time you play a C chord? Or press the button to your game system with a solenoid when you play the note E3? There are a lot of ways to interpret this project for your own unique use cases. ⊘

DR. SQUIGGLES AI RHYTHM ROBOT

Written and photographed
by Michael Krzyzaniak

BUILD A SMART OCTOPUS DRUMBOT THAT LISTENS, LEARNS, AND PLAYS ALONG WITH YOU

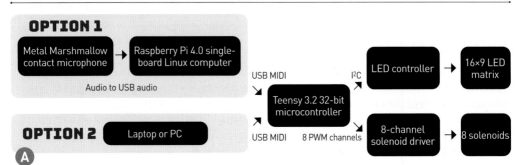

OPTION 1

Metal Marshmallow contact microphone → Raspberry Pi 4.0 single-board Linux computer

Audio to USB audio

USB MIDI

Teensy 3.2 32-bit microcontroller

I²C → LED controller → 16×9 LED matrix

8 PWM channels → 8-channel solenoid driver → 8 solenoids

OPTION 2   Laptop or PC

USB MIDI

A

Have you ever wanted your own cute and friendly robot that you can play music with? Dr. Squiggles is just that — a musical drumming robot that plays rhythms by tapping on whatever surface you set it on. Moreover, if you tap rhythms, it can listen to you through its microphone, synchronize to your beat, imitate you, or learn to play rhythms that are similar but not identical to yours!

When I designed Dr. Squiggles, I wanted to make the most mechanically simple robot I could think of, and try to imbue it with the most complex behavior possible. Essentially, it's just a contact microphone and eight solenoids controlled by an embedded computer.

A functional block diagram of Dr. Squiggles is shown in Figure Ⓐ:

- Eight **solenoids** tap the rhythms.
- A low-res **LED matrix** serves as the robot's eye.
- A **contact microphone** listens selectively to rhythms that you play.
- An embedded **Raspberry Pi computer** does the "smart" tasks of analyzing what you play and generating rhythms.
- A **Teensy microcontroller** animates the eye and receives USB-MIDI messages from the Raspberry Pi, upon which it applies voltage to the solenoids, causing them to eject and make tapping sounds.

NOTE: You can also build a MIDI-only version of Dr. Squiggles, without the Raspberry Pi and contact microphone. Connect the Teensy to a computer and use your favorite music software or programming language to send MIDI messages for the robot to play. And see **Glocken' Roll** on page 84 for a wireless BLE MIDI instrument that also uses tapping solenoids, but for melodies.

I designed Dr. Squiggles as part of my research at University of Oslo, where I study human-robot and robot-robot interaction in the context of music. Recently I've been especially interested in how swarms of simple robots can make music together, so my colleagues and I built ten Dr. Squiggleses. We got really good at building them and so we wanted to share the process with you!

TIME REQUIRED:
Several Weekends

DIFFICULTY:
Advanced

COST:
$400–$500

MATERIALS

- » **Raspberry Pi 4 single-board computer, 2GB RAM**
- » **microSD card, 16GB high speed** Mouser #922-605510
- » **Teensy 3.2 microcontroller** Adafruit #2756, adafruit.com
- » **Contact microphone** Metal Marshmallow piezo disc contact mic and preamp. I make and sell these at etsy.com/shop/MichaelKrzyzaniak.
- » **Push-pull solenoids, 12VDC (8)** Adafruit #412
- » **Squiggles Controller printed circuit board (PCB)** oshpark.com/shared_projects/FstH6dIJ, or just use perf board, Adafruit #2670
- » **Charlieplexed LED Matrix, 16×9** Adafruit #2974
- » **LED driver, IS31FL3731 with header pins,** Adafruit #2946
- » **Plywood sheet, 4mm (⁵⁄₃₂") thick, 300mm×400mm** 3.75mm measured thickness
- » **Clear spray acrylic**
- » **Furniture paint, black semi-gloss**
- » **Glitter**
- » **Acrylic sheet, clear, 2mm×40mm×60mm**
- » **Step-down voltage regulator, switching, 2.5A 5V** Pololu #2858, pololu.com
- » **Darlington driver IC chip, ULN2803** Adafruit #970
- » **Screw terminal block** Mouser #490-TB007-508-02BE, mouser.com
- » **Header pins, single row** Mouser #571-9-146278-0
- » **Header pins, double row** Mouser #571-5-146254-8
- » **Molex female crimp terminals (32)** Mouser #538-90119-2109
- » **Molex sockets: 2-position (11), 4-position (2), and 1-position (2)** Mouser # 538-90123-0102 and 538-90123-0104; Pololu #1900
- » **Switch, momentary pushbutton** Mouser #633-SB4011NOH
- » **Switch, on/off toggle** Mouser #633-M2011SA1W01
- » **USB cable, 8"** Amazon #B075ZQKH63
- » **DC power supply, 12V 5A** such as RND Power #320-00034, Distrelec #30098154
- » **Hobby felt, 1mm**
- » **Machine screws, M3×8mm (16)** with nuts and washers
- » **Yarn, 1 skein (optional)** plus a little in a contrasting color, for knitting the balaclava. You can also modify a regular balaclava, or come up with your own covering for Dr. Squiggles.

CONTINUED ON THE FOLLOWING PAGE

Adobe Stock - dlyastokiv

» ¼" guitar cable
» ¼" to 3.5mm audio adapter Amazon #B000068O47
» **USB audio dongle** Amazon #B07RV6VBNR
» **USB extension cable, 3m/10'** microphone to Pi, RND #765-00063, Distrelec #30125781
» **Sandpaper, 300–600 grit**
» **Gorilla 2-part epoxy**
» **Wood glue**
» **Small wires**
» **Super glue gel**

TOOLS
» **Utility knife**
» **Laser cutter**
» **Paintbrush**
» **Computer**
» **Needle or small blade**
» **Hacksaw and dust mask (optional)** if you're cutting perf board
» **Soldering iron and solder** flux optional
» **Hot glue gun**
» **Molex crimping tool**
» **Screwdriver and small wrench**
» **Knitting needles or Addi Express king size circular knitting machine (optional)** if you're knitting the balaclava, Amazon #B004HS7T7S
» **Yarn needle (optional)**
» **Crochet hook, about 4.5mm (optional)**

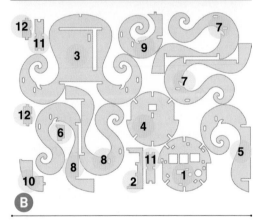

B

Cut

C

BUILD YOUR DR. SQUIGGLES RHYTHM ROBOT

1. CUT AND PAINT PLYWOOD PARTS
Start by laser-cutting the plywood parts for Dr. Squiggles' body (Figure **B**). Download the OpenSquiggles repository at github.com/michaelkrzyzaniak/OpenSquiggles, then follow the cutting and painting instructions on the project page, makezine.com/go/dr-squiggles.

2. PROGRAM THE TEENSY
Start by installing the Teensyduino software on your laptop or PC, from pjrc.com/teensy/td_download.html. Then replace one of Teensyduino's native files with a file from the OpenSquiggles repository. The file you're replacing is called *usb_desc.h*. On OSX, it's found in */Applications/Arduino.app/Contents/Java/hardware/teensy/avr/cores/teensy3/usb_desc.h*. Replace it with the file found in the OpenSquiggles repository at *OpenSquiggles/Firmware/Teensy_Replacements/usb_desc.h*. This will make the Teensy enumerate as a USB MIDI device called Dr Squiggles and not Teensy MIDI, so that the OpenSquiggles main software will talk to it properly.

Now find Dr. Squiggles' firmware in the repository at *OpenSquiggles/Firmware/Firmware.ino*, and open it using the Teensyduino software. From the main menu, select Tools → Board → Teensy3.2 / 3.1, and then Tools → USB Type → MIDI. Connect your computer to the Teensy via USB. Press the Upload button in Teensyduino to program the Teensy, then disconnect the Teensy from your computer.

3. CUT THE TEENSY VIN TO VUSB TRACE
Use a needle or small blade to sever the tiny trace on the bottom of the Teensy that connects USB power to the rest of the board (Figure **C**). After you've severed it, you can use a multimeter to check that there is no longer continuity between the two pads.

This makes it safe to power the board with an external power supply, but it means the board can no longer be powered via USB. To be reprogrammed it will need an external power supply, which won't be added until Step 9.

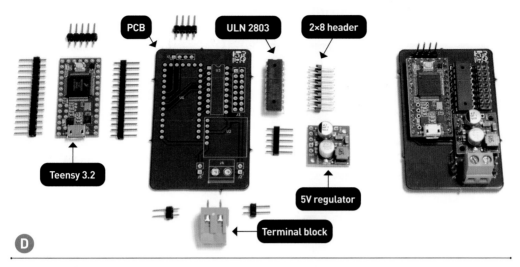

PCB

ULN 2803

2×8 header

Teensy 3.2

5V regulator

Terminal block

D

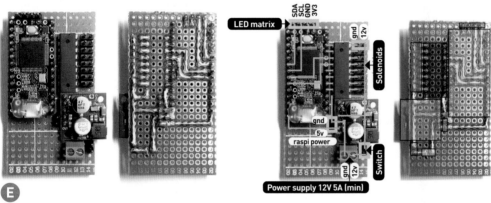

SDA
SCL
GND
3V3

LED matrix

gnd
12v

Solenoids

gnd
5v
raspi power

Switch

gnd
12v

Power supply 12V 5A (min)

E

4. ASSEMBLE THE MAIN CONTROLLER BOARD

The main controller circuit controls Dr. Squiggles' solenoids and eye, and supplies power to the Raspberry Pi. This module has two variants.

- **PCB version:** Download the PCB design file *OpenSquiggles/Hardware/Board.zip* and have it manufactured at a PCB house. Or order PCBs direct from OSH Park at oshpark.com/shared_projects/FstH6dlJ (3 board minimum).

   Solder header pins to the Teensy and the voltage regulator. Then solder the Teensy, regulator, Darlington driver, terminal block, and remaining headers to the PCB (Figure **D**).

- **Perf board version:** Use a hacksaw to cut a piece of generic perf board 13 holes wide and 23 holes tall. Solder header pins to the Teensy and the voltage regulator. Tack-solder the parts as shown in Figure **E**, and then connect them by bridging adjacent pads on the bottom

of the board with solder. The right-hand side of Figure E shows the finished board with the trace locations drawn over the top (component-side) of the board, and the component locations drawn over the bottom (solder-side) of the board, which will help you during assembly.

5. SOLDER THE LED MATRIX TO ITS DRIVER

Now you'll prepare the electronics for Dr. Squiggles' eye module. Solder a 7-pin header to the top of the LED driver board (IS31FL3731), as shown in Figure **F**. Then use two 13-pin headers to solder the LED driver back-to-back to the LED matrix. The LED matrix has rotational symmetry, so it doesn't matter which way you install it, as long as the LEDs are facing out. Adafruit has a

F

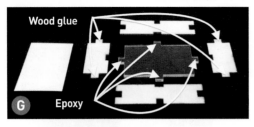

detailed assembly guide at learn.adafruit.com/i31fl3731-16x9-charliplexed-pwm-led-driver.

6. MAKE THE EYE MODULE

Cut the acrylic and paper parts following the instructions at makezine.com/go/dr-squiggles. Glue the four plywood pieces labeled 11 and 12 (two of each) into a rectangle (Figure **G**). Next, mix up a small amount of epoxy, put a drop on each of the four tabs on the acrylic panel, and then press the tabs into their slots in the plywood rectangle. Wait for everything to dry.

Drop the paper diffuser into the box so it's lying on the acrylic. Then drop the LED assembly in so the LEDs are pressed up against the paper (Figure **H**). Put a dab of hot glue in one or two of the inside corners, to lightly pin the LED driver board inside so it can't fall out.

7. START GLUING THE BODY

Insert the Raspberry Pi USB and Ethernet jacks into Part 1, the bottom plate (Figure **I**). Clip them in place by gluing Part 2 (Pi clip) to Part 1. Slide the bottom slit of Part 3 (left and right tentacle plate) over Part 1, and glue 3 to 2 and 1, as shown.

Now glue Part 4 (top plate) into the top slit of 3, so its holes extend over the Raspberry Pi SD card slot (Figure **J**). Glue Part 5 (rear tentacle) into the rear slots on 4 and 1. Glue Part 6 (rear tentacle upper extension) to 3 and 4.

Glue the two Parts 7 (rear-left and rear-right tentacles) into 3 and 4. Take a break from gluing.

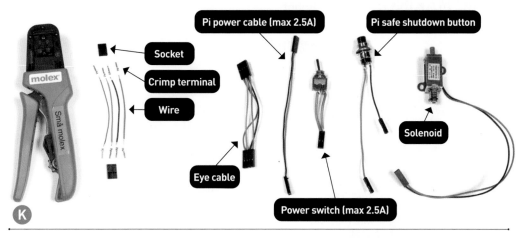

Pi power cable (max 2.5A)

Pi safe shutdown button

Socket

Crimp terminal

Wire

Solenoid

Eye cable

Power switch (max 2.5A)

K

Board

Switch

Button

USB

Power

L

8. MAKE CONNECTOR CABLES

Start by making the eye module cable. Cut four 7cm lengths of wire and use a Molex crimping tool to attach a female crimp terminal to each end. Then insert the crimped wire ends into plastic 4-position crimp housings. By the same process, make the 2-wire, 12cm Pi power cable.

Also solder two 2cm wires to the on/off toggle switch, and crimp a 2-position terminal to the free ends. Do the same to the pushbutton using a 5cm and 10cm wire. The toggle is Dr. Squiggles' on/off power switch. The pushbutton is a safe-shutdown button for the Raspberry Pi.

You also need to crimp a 2-position connector to the end of all the solenoids, however, I recommend waiting until after Step 12 so you can trim the solenoid wires to length first. All the completed cables are shown in Figure **K**.

9. INSTALL THE ELECTRONICS

Use plenty of hot glue to affix the main controller board to Part 3 (Figure **L**). Mount the toggle and pushbutton in their holes in Part 1 using the nuts and washers that came with them.

If your power supply came with a connector on the end, cut it off and discard it, and strip and tin

a few millimeters at the end of the wires. Thread this power cable through its hole in Part 1, put the tinned ends into the screw terminals on the main controller board, and screw them in place. Hot-glue these wires in place to prevent them from being accidentally pulled out later. (The correct polarity of the power cable in the terminal is shown in Figure **P** on the following page).

Also thread the micro USB cable through its hole in Part 1. Plug it into the Teensy and secure it from below with lots of hot glue, because this USB jack is prone to breaking off irreparably.

10. FINISH GLUING THE BODY

Glue Part 9 (front tentacle) to 3, then glue Part 10 (front tentacle upper extension) to 4. Finally, slide the two Parts 8 (front-left and front-right tentacles) into 3 and 4 (Figure **M** on the following page), but leave these unglued so the circuitry can be accessed.

11. DAMPEN THE SOLENOIDS

When voltage is applied to a solenoid, it ejects and, in our case, strikes something, making a sound. Then, a short time later, when the voltage is removed, the solenoid retracts, and the plunger

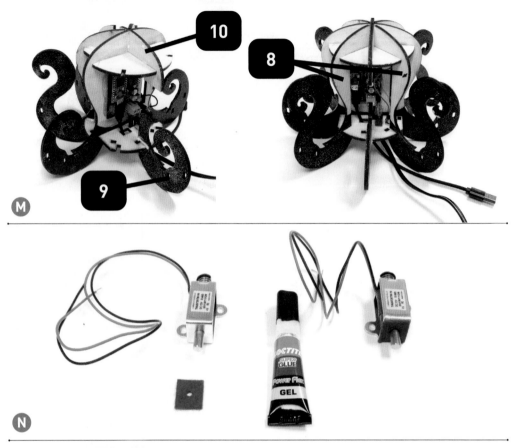

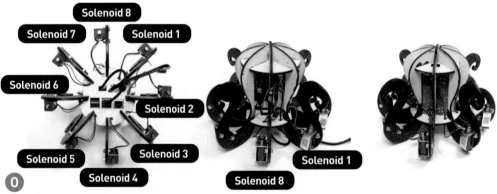

rams into the metal frame of the solenoid, making a second sound. This second sound is undesirable, and in this step you will dampen it.

Cut a small rectangle out of 1mm thick hobby felt. Cut a small hole in the center of the felt, and stretch it over the solenoid plunger. Use gel cyanoacrylate super glue to glue the felt to the frame of the solenoid (Figure **N**), so that when

the plunger retracts it strikes the felt instead of the metal frame. Do this to all 8 solenoids.

12. MOUNT THE SOLENOIDS

Screw a solenoid onto each tentacle, using a washer and a nut on the back of each screw. Carefully adjust each solenoid up and down until it just barely strikes the table when fully ejected,

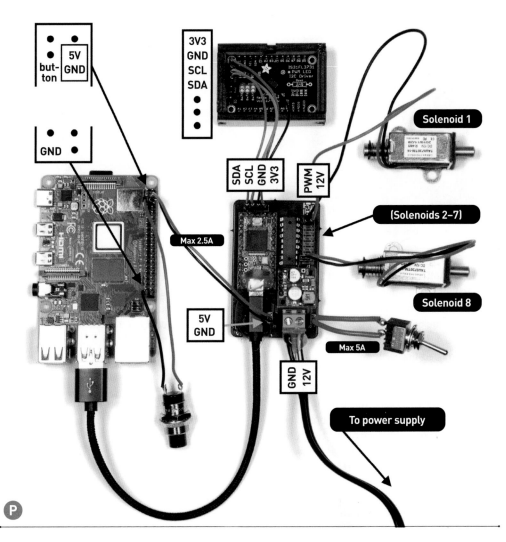

5V
but-
ton
GND

3V3
GND
SCL
SDA

GND

SDA
SCL
GND
3V3

PWM
12V

Solenoid 1

(Solenoids 2–7)

Max 2.5A

5V
GND

Solenoid 8

Max 5A

GND
12V

To power supply

P

and then tighten the nuts.

Thread the solenoid wires through their holes in the bottom plate, Part 3 (Figure O).The four wires from the back of the robot can be threaded between the back of the Raspberry Pi and the tentacles, and then passed to the front of the robot through the small remaining hole in Part 3, just next to the Pi headers. Now you can trim the solenoid wires to length and crimp connectors to the ends, as in Step 8.

13. WIRE EVERYTHING TOGETHER

Connect all the circuit components using the cables you made in Step 8, following the wiring diagram (Figure P). Of course you'll connect all these components in place inside the body, not out in the open as shown.

14. GLUE THE EYE MODULE TO THE BODY

You've already connected the eye module to the main controller board. Now glue it to Parts 9 and 10 (front tentacle lower and upper parts). Make sure the row of header pins that connect the eye to the controller are on your left when you're face to face with the robot (Figure Q), otherwise the eye will be upside down.

15. KNIT A BALACLAVA HAT (OPTIONAL)

I used an Addi Express king size circular knitting machine, although a skilled knitter could do this by hand. You can follow the complete instructions on the project page at makezine.com/go/dr-squiggles. You could also just buy a small balaclava and modify it to fit the eye, or improvise a different covering for your Dr. Squiggles.

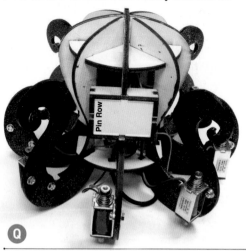

Pin Row

Notch

Q

R

16. PROGRAM THE RASPBERRY PI SD CARD

The Raspberry Pi OS (formerly Raspbian) requires a good amount of configuration to work in Dr. Squiggles, so to simplify installation, I've made a disk image that contains the fully configured operating system. Download it onto your computer from michaelkrzyzaniak.com/squiggles/os and unzip it. Copy the disk image onto a 16GB microSD card, using the following tutorial for MacOS (makezine.com/go/pi-clone-macos), or whatever works for your computer.

Once you're done, remove the card from your computer and insert it into the Raspberry Pi's card slot.

17. PUT THE BALACLAVA ON

Each tentacle has a notch to hold the balaclava (Figure R). The balaclava is 24 stitches around, so every third chain stitch hooks onto a tentacle.

18. CONNECT THE MICROPHONE

Because the Raspberry Pi's 3.5mm jack only has audio out (headphones) and not audio in (microphone), you'll need to use an audio-to-USB converter. This requires several adapters. I used the following sequence (Figure S):

- Contact microphone
- ¼" guitar cable
- ¼" female to 3.5mm male mono audio adapter
- 3.5mm female audio to USB audio converter
- USB extension cable
- Raspberry Pi

NOTE: The extension cable should terminate in a short plug, as there is not much clearance underneath the robot for a long plug.

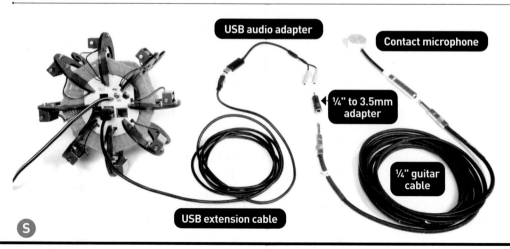

USB audio adapter

Contact microphone

¼" to 3.5mm adapter

¼" guitar cable

USB extension cable

S

MICHAEL KRZYZANIAK builds robots that play music. He works as a researcher at the RITMO Centre for Interdisciplinary Studies in Rhythm, Time, and Motion, University of Oslo.

Annica Thomsson

START YOUR SMART SYNCOPATED CEPHALOPOD

When you start up the robot, the Raspberry Pi waits for you to enter some commands over Wi-Fi, or from a computer using SSH. To do this, first use a router, computer, or your phone to create an access-point network (Wi-Fi hotspot) with the following credentials:

Network Name: Dr_Squiggles
Password: IsAwesome!

Plug in the robot, turn it on using the toggle switch, and wait a minute for it to boot. It will automatically connect to the network, download the most recent OpenSquiggles software, and compile it. Connect a computer to the network you just created and open a terminal emulator (such as Terminal.app on OSX). Enter the command:

`ssh pi@dr-squiggles.local`

and when prompted, enter the password

`drsquiggles`

You should see

`pi@dr-squiggles:~ $`

You're now controlling Dr. Squiggles' Raspberry Pi. (You should change both passwords, and optionally the network name, using the **raspiconfig** command.) To run the OpenSquiggles software, enter the command **sq** .

PERCUSSIVE PERMUTATIONS

Tap out some rhythms on the table and Dr. Squiggles will start playing with you!

- First he listens through the contact mic to learn your rhythm. It takes a moment to synchronize. You'll see him blink his eye to the beat.

- Now he'll use his interactive rhythm generation algorithms and his 8 solenoid tappers to embellish your rhythm or complement it.
- Play a new rhythm and he'll learn that one and adapt again!

When you're done jamming, press and hold the button on the bottom of Dr. Squiggles for at least 2 seconds. This will safely shut down the Raspberry Pi. Wait about 30 seconds and then shut off the power using the switch.

GOING FURTHER

Dr. Squiggles can play rhythms on — and with — just about anything! We've connected Dr. Squiggles to drums and other percussive instruments, triggered him with myoelectric muscle sensor arm bands, and daisy-chained an ensemble of Dr. Squiggleses to learn rhythms from each other.

And of course you can edit his code to change his beat behavior or his eye animations to suit your project. We hope you'll share your experiences on the project page at makezine. com/go/dr-squiggles. ◑

Thanks to Habibur Rahman, who helped with building and did all the nicest work shown here, and to Kyrre Glette for all his support and feedback. This work was partially supported by the Research Council of Norway through its Centres of Excellence

Watch Dr. Squiggles play music with people and other Dr. Squiggleses at youtu.be/yN711HXPfuY and youtu.be/ohH1TN1MRYw, and learn how they interact with each other at ambisynth. blogspot.com/2020/08/breadth-first-binary-numbers-and-their.html.

PRINT-A-SYNTH

**Written and photographed
by Johan von Konow**

GET STARTED WITH DIY SYNTHS WITH THIS 3D-PRINTABLE, NO-PCB KEYBOARD

If you want to play with MIDI, complement your music equipment with a pocket keyboard, or learn electronics, this project is for you. LEET is my vision of a new kind of modular affordable synth, complete with vibrant blinkenlights, that you 3D print and build yourself.

Along with the keyboard, I have designed a drum pad, chord keyboard, control unit, arpeggiator, and a step sequencer (Figure Ⓐ). One special feature is that the units have RGB LEDs for each key, enabling playback visualization (so each device is both MIDI out and in). This is helpful for music training and editing, but it also looks great. They can be used as input devices to any computer with a DAW (Digital Audio Workstation) like Ableton, Logic, Cubase, GarageBand, etc. They run on Windows, Mac, or Linux (including Raspberry Pi). They can even be connected to your mobile phone (Android or iOS), providing a tactile, super portable music development platform (Figure Ⓑ).

The LEET devices are built around a single 3D-printed core (3DPCB) that creates the exterior of the product, holds all components in place, and has integrated wire channels that connect the components in a foolproof way. 3DPCB makes the devices easier to replicate, less expensive, and you don't need to wait for PCB delivery.

Everything is open source at vonkonow.com (Figure Ⓒ) and can be tweaked, improved, and complemented with new functions to fit different needs. Build a wooden stand, add a light sensor to control the timbre, connect to a Speak & Spell, or build something completely new!

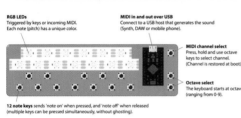

RGB LEDs
Triggered by keys or incoming MIDI.
Each note (pitch) has a unique color.

MIDI in and out over USB
Connect to a USB host that generates the sound
(Synth, DAW or mobile phone).

MIDI channel select
Press, hold and use octave
keys to select channel.
(Channel is restored at boot).

Octave select
The keyboard starts at octave 4
(ranging from 0-9).

12 note keys sends 'note on' when pressed, and 'note off' when released
(multiple keys can be pressed simultaneously, without ghosting).

The velocity is fixed to 50% (can be changed in firmware).
There is no aftertouch, nor pitch bend in this version.

Dimensions: 154x45x8mm (6x1.8x0.3 inches).

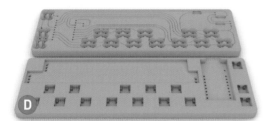

TIME REQUIRED:
An Afternoon

DIFFICULTY:
Moderate

COST:
$6–$10

MATERIALS
» **3D-printed core** I call it the 3DPCB.
» **Arduino Pro Micro microcontroller board, 5V, ATmega32U4** or compatible clone
» **Tactile switches, 6×6mm, 4 pin through-hole (15)** preferably with low activation force: 50g/0.5N
» **LED strip, WS2812, 60 LEDs** I used IP60, white FPC; you need only 13 LEDs, 22cm).
» **Bare copper wire, approx. 0.3mm diameter, 10" lengths (24)** AWG 28 or 29. I stripped a stranded cable to get mine.

TOOLS
» **3D printer, FFF/FDM type** with PLA filament
» **Soldering iron with narrow tip**
» **Pliers: needlenose and cutting**
» **Hobby knife**
» **Hot glue gun**
» **Computer with Arduino IDE software** free from arduino.cc/downloads
» **Micro-USB cable**

JOHAN VON KONOW is a creative engineer with knowledge in mechanics, electronics, software, and industrial design. Based in Stockholm, Sweden.

PRINT THE 3DPCB

Print the core using a standard filament-based 3D printer. I recommend starting with the smaller *leet_test* part to verify that your printer is calibrated before embarking on the 3- to 4-hour-long print (Figure D).

I used a Prusa i3 MK3S with 0.4mm nozzle and the following settings:
• **Material:** PLA @ 210°C
• **Layer height:** 0.2mm
• **Infill:** 20%
• **Support:** None

PROGRAM THE ARDUINO

While printing you can download, set up, and program the microcontroller:

1. Install the Arduino IDE and download the latest LEET keyboard firmware from github.com/vonkonow/LEET-Synthesizer/tree/main/Keyboard.
2. Download and install the libraries for MIDIUSB (github.com/arduino-libraries/MIDIUSB) and NeoPixelBrightnessBus (github.com/Makuna/NeoPixelBus).
3. Connect the Arduino to your computer with a Micro-USB cable.
4. Select Tools → Boards → Arduino Pro Micro, and select the corresponding COM port.
5. Compile and upload the code.

PREPARE THE WIRES

Cut 24 pieces of bare 0.3mm copper wire 10" long. I stripped some "RK wire," a stranded cable

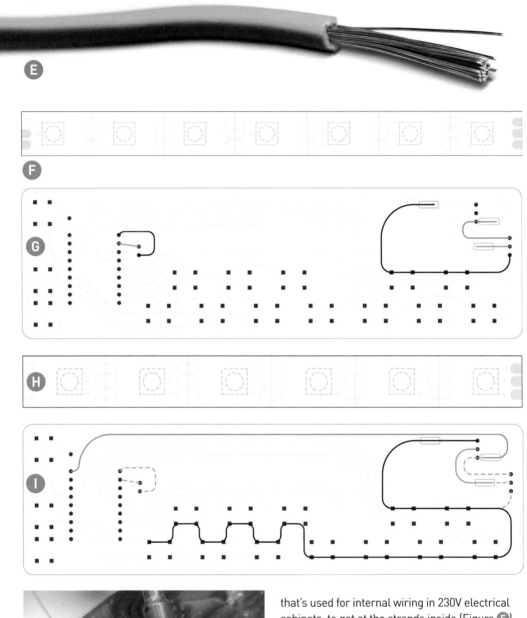

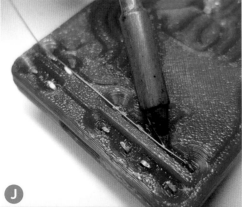

that's used for internal wiring in 230V electrical cabinets, to get at the strands inside (Figure **E**). In the U.S., you could look for something like 14 AWG cable with 28 gauge strands, or 15 AWG with 29 gauge strands. (The last number is most important — the diameter of the strands.) But you can also use plain bare wire (not coated "magnet wire") approximately 0.3mm in diameter.

ATTACH THE ELECTRONICS

1. Secure the Arduino Pro Micro (without pin headers) in place with a dab of hot glue on the

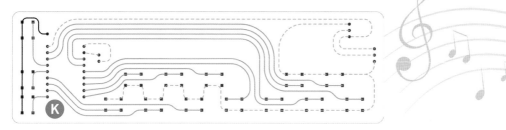

backside. Ensure that the holes are aligned.

2. Next, insert the switches and fold the pins on the backside to lock them in place. I recommend using a small flat screwdriver.

3. After that, cut a section of LED strip with 7 LEDs (Figure **F**). On the back, remove the protective liner and the double-sided adhesive that covers the three connections on each end.

4. Solder two strands of wire to the exposed DIN and GND pads on the backside (note that one end of the LED strip is DIN and the other is DO). Solder three strands of wire to the other end (DO, 5V, and GND).

5. Thread all 5 wires through the 3DPCB (Figure **G**), remove the liner, and press the LED strip into place.

6. Thread the DIN wire through the wire channel to the microcontroller, ensure it's tight, and solder it in place on the front side. Cut away excess wire close to the solder.

7. Thread the neighboring GND through the channel, stretch, solder it to the Arduino board, and trim.

8. Thread the three remaining wires on the other end through their channels to each dedicated opening.

9. Now prepare another section of LED strip, this time with 6 LEDs.

10. Remove the protective liner and the double-sided adhesive that covers the three connections on the DIN side (Figure **H**).

11. Solder three strands of wire to the exposed DIN, GND, and 5V pads on the backside.

12. Thread the 3 wires through the 3DPCB, remove the liner and press the LED strip in place (Figure **I**).

13. Thread the 5V wire through the channel, stretch, and solder it with the wire from the other LED strip in the dedicated opening. Trim the wire from the other LED strip, then continue to thread it to the microcontroller, stretch it tight, and solder in place.

14. Thread the DIN wire through the wire channel to the opening where it meets the DO wire from the other LED strip. Ensure both wires are tight, solder them together, and trim the ends.

15. Thread the GND wire and solder it to the other GND wire at the dedicated opening. Trim the wire from the other LED strip and continue to thread this wire through its channel to each of the 12 keyboard switches. Tighten, and solder the wire to each switch (24 pins) along the way.

CONNECT THE SWITCHES

Solder a wire to both available pins of a switch. Insert it in its wire channel, thread it through the Arduino, add some tension, solder it, and cut the excess (Figure **J**). Repeat for all 15 switches (Figure **K**). Don't forget to take a break every now and then to stay focused.

INSPECT

Check the solder joints, and ensure that each wire is secure and tight within its channel and that everything is flush against the backside. If anything is loose, protrudes, looks funny or sharp, re-solder until it's perfect.

Good job — give yourself a well-deserved high-five.

USE IT!

Connect the keyboard to a computer or a mobile phone with a USB micro cable. The startup animation should turn on every LED, and the device should appear as MIDI in your DAW. Insert a coin and start playing!

Check out the other LEET devices at vonkonow.com, especially the sequencer that allows recording and playback of songs from your keyboard. This is also the place to share your build in a forum, as well as exchange ideas, assist with development, and get support or help from other LEET builders. ◗

GO LONG!

Need to monitor remote sensors on your next project? **LoRa** might be your best wireless bet. **Written by Andreas Spiess**

DON'T MIND ME. I'M JUST HAVING A LITTLE WANDER AROUND.

LoRa is a long-range IoT protocol commonly used for industrial applications — and it's the most common for makers too. It's a free-to-use system that allows for small messages to be wirelessly transmitted over miles of distance under optimal conditions. LoRa is very good with sensor networks if these sensors do not transmit a lot of data. For example, it's perfect for sending humidity data for plants (Figure **A**), because the soil humidity usually does not change in seconds. Or for tracking animals' positions on large farms (Figure **B**) and game reserves, because animals don't move too fast. Or for monitoring parking lot occupancy (Figure **C**) where you don't expect too many changes per day.

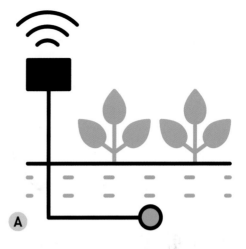

HOW LORA WORKS

LoRa is a *low-power, wide-area network* standard, also called *LPWAN*. This term consists of three parts and they're all important.

1. Network

The difference between a typical small device and an Internet-of-Things (IoT) device is its ability to connect to the internet. And because we expect millions of them, we need a network to connect all of them. This network must be based on standards because different companies will build the network and the IoT devices. Best is always an international standard accepted by "everybody."

2. Wide Area

Our ESP8266 or ESP32 devices can connect to a Wi-Fi network, which is part of a *LAN*, or *local area network*. We all know that its reach is limited to a few meters around the access points. *Wide area networks* can bridge much bigger distances. This is necessary for IoT devices because we want to use them everywhere.

Old-timers can remember receiving AM radio in the middle of nowhere, far away from the station. This was really "wide area." But the transmitters were usually huge and able to emit kilowatts of energy. It's relatively easy to bridge long distances using high power. This brings us to the third term.

3. Low Power

If we want devices to work on batteries, we don't have a lot of power for transmission. And now we

ANDREAS SPIESS runs a large YouTube channel focused on the "Advanced Maker." He has loved wireless technologies ever since he became a HAM radio operator in 1977 (HB9BLA).

Adobe Stock (HN Works, DrawWing.Time, sabuhinovruzov, talsen, 06photo), Bosch

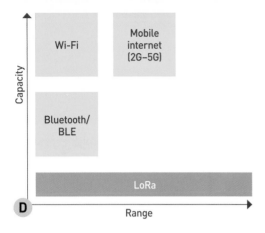

D

see the dilemma: We want miles of reach but we have no power to spend.

Fortunately, physics gives us a third parameter to ease this dilemma a bit: *Bandwidth*. Physical laws say that if we want to create a wireless connection across a certain distance, we can either increase transmission power or decrease the channel's bandwidth.

But why should we bother with bandwidth? Because bandwidth and the maximum *capacity* of a channel are directly related: The smaller the bandwidth, the lower your channel's capacity. I still remember the old days of Morse communication, where a good operator could transmit two characters per second, which is a little less than 20 bits/second. Today, our wireless LANs are capable of transferring millions of characters per second. And they are still always too slow.

COMPARING WIRELESS PROTOCOLS

We can use the chart in Figure **D** to visualize the relationship between capacity and range. On the x-axis, we have the range, and on the y-axis, the capacity. Let's look at some of the well-known technologies and where they fit.

Wi-Fi has a high capacity but only a low reach. And we know from our ESP8266, it is quite power-hungry. Not in the kilowatts as the old radio stations, but it quickly needs a quarter of a watt during transmission.

Next is the *mobile internet* on our smartphones. The reach here is hundreds of feet up to a mile or so in rural areas. Capacity is relatively high. But also here, you don't get fast 4G or 5G coverage if you're in the middle of nowhere, because the next antenna tower is probably over a mile away. And we all know that our smartphones' battery life is not great — this technology needs quite a bit of power.

Bluetooth, next, has less capacity than Wi-Fi. As we know from many gadgets, it runs well on small batteries. But, unfortunately, its reach is only a few meters. So, none of these technologies fulfill our IoT devices' needs for low power *and* wide area.

Then we have *LoRa*. It has its own space: Long range, but, because of power limitations, also low bandwidth. And this bandwidth is limited not only by the law of physics, but also by human laws, as we will later see.

So, we now know where LoRa fits. It is not comparable with Wi-Fi and not at all a

DISTANCE TESTING AND RESULTS

While under optimal everyday conditions you might expect a LoRa range reaching 1 mile or more, I undertook an experiment to establish my longest successful transmission. By utilizing a high-altitude gateway in Weissenstein, Switzerland, and a comparable location in Germany, I was able to transmit a signal just over 200km (about 125 miles). You can see this in my video: youtube.com/watch?v=adhWlo-7gr4

COMPETING TECHNOLOGIES

Other new commercial networks provide long-range communication options for IoT. **Sigfox** offers broad geographical coverage and can be used for maker projects. LTE-based cellular networks like **Cat-1**, **Cat-M**, or **NB-LTE** are emerging, but the modules are expensive and not easy to use because of a lack of information. We hope to find more projects in the future to learn from.

These types of commercial networks, of course, require paid service contracts.

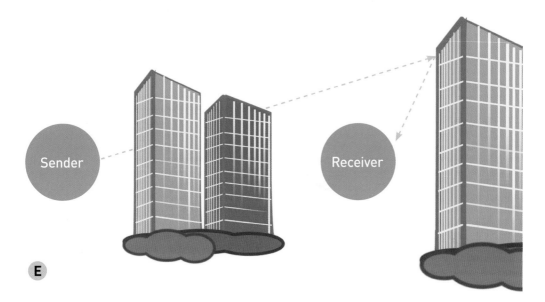

E

LINK BUDGETS

To better understand the "range" of a wireless connection, we have to deal with a relatively complex topic: The *link budget*. What is this, and why is it so important?

Like every other budget, the link budget is something you have initially and which you spend over time. If your budget is used up, you cannot spend more.

The link budget has to do with the link between the transmitter and the receiver. It is filled by the transmitter's transmission power and the receiver's sensitivity and is calculated in decibels or dB. The link budget is deducted by all sorts of obstacles between the transmitter and the receiver like cables, distance, walls, trees, etc. If the link budget is used up, the receiver will not produce any usable signal.

So, what is the link budget for LoRa compared with other technologies like LTE or 4G? LoRa has a link budget of 154dB, which is much higher than mobile internet's 130dB, or Wi-Fi at around 100dB.

Let's look at what this means by doing some calculations: First, assume we have a line-of-sight connection between the transmitter and

the receiver. Our LTE budget of 130dB would be good to bridge around 80km. LoRa's link budget of 154dB on the same frequency is good for 1,300km! You see, it matters a lot.

This is also theoretical. If we insert 5 meters of thin cable between our transmitter or receiver and antennas, we lose about 8dB. This reduces LoRa's maximum distance to 500km. So, 10 meters of cable is equivalent to 800km in free air.

Next, we have to spend parts of our budget on obstacles like walls or trees between the transmitter and receiver. The thicker and the more conductive the obstacle, the more budget it eats. And sometimes we don't have a line of sight, so we have to work with reflected signals, which reduce the link budget immensely (Figure E).

LoRa achieves a link budget higher than LTE not because of technological breakthroughs, but simply because of its very narrow bandwidth and a low data rate. Its rated capacity ranges from 250 bit/s to 250 kb/s. That's very low compared with the megabits of LTE.

CONNECTING TO LoRaWAN

Now let's talk about the network that's needed between LoRa devices and our IoT applications. This is where *LoRaWAN* comes into play. It consists of distributed *gateways* connected to the internet and an infrastructure capable of relaying the IoT messages to our applications.

LoRaWAN INFRASTRUCTURE OVERVIEW

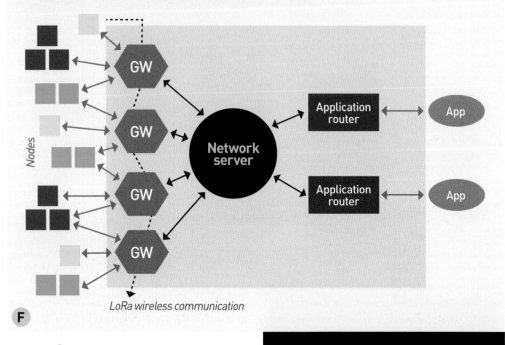

LoRa wireless communication

F

Figure **F** shows an overview of the whole infrastructure, with many devices or **nodes** connected to one gateway. Many gateways are connected to a **broker** network server infrastructure, and many applications are also connected to the same brokers.

Two network approaches exist — commercial and community-based. In many places, telecom companies have started to deploy private LoRa networks. As with cellular phones, you can buy a contract and use this infrastructure. You just have to connect your LoRaWAN-enabled nodes to the network. Such networks exist in the Netherlands and Switzerland, for example.

The community approach is led by The Things Network (TTN). This Dutch group built an infrastructure to transfer the messages between the gateways and the applications. As a startup, they needed gateways all over the world but they had no money to invest, so they asked people like you and me to build such a gateway and deploy it. They provide a map of all available gateways, where you can check if one is close to you (see page 108). If so, you only have to switch your node on, and it is connected to the TTN network. Like a commercial service, but free of charge.

FREQUENCY MATTERS

LoRa networks work on particular **ISM frequencies** reserved for industrial, scientific, and medical usage. Please consult the specifications and laws for your country before you purchase an IoT device. LoRa usually runs on **868MHz in Europe**, on **915MHz in the U.S.**, and **433MHz for certain Asian countries**. You can learn more at lora-alliance.org/lorawan-for-developers.

All these frequencies have something in common: They are **free bands**, and you don't need to apply for a license or pay a fee to use them. Very attractive! But this comes with a handicap: The allowed power is only 25mW in Europe and 1 watt in the U.S. Not a lot, if you consider that amateur radio operators are allowed to emit up to 1,500 watts, for example. But it's perfect for battery operations.

BUILDING YOUR OWN DEVICES

If there is no gateway where you are, or you don't want to depend on somebody else's, you can build your own. An easy version consists of a concentrator PCB and a Raspberry Pi. Because the concentrator has 8 channels, it can support up to 8 IoT devices in parallel. Not a lot if we think about the numbers of projected IoT devices. So, what to do?

If we agree that each device will only use its channel, let's say, for 50% of the time, then one gateway could already support 16 devices. And if each device will only use the channel 1% of the time, a gateway could support 800 devices (Figure **G**). That is precisely the concept of LoRaWAN. And this concept is also in line with the law. European governments typically allow one node or gateway device to transmit only 1% of the time. This is known as a *duty cycle* of 1%.

Because of that, we have to divide the 250 bps transmission speed by 100. This results in the worst case of 2.5 bps — slower than Morse (Figure **H**). Unfortunately, we're not yet finished with reducing capacity: If two nodes transmit simultaneously on the same frequency, there is a collision, and the messages are lost. So if we deploy too many devices in one area — generally, over 1,000 nodes — the gateway capacity is reduced because of such collisions. (LoRa networks, mainly built for sensors, favor the direction from the sensor to the gateway. The traffic in the other direction is limited.)

A sensor node consists of at least three components: a sensor, a microprocessor, and a communication module. To build one, you can use your microprocessor of choice and connect it with a LoRa module.

I suggest using either a ready-made sensor node or a module from the company TTGO for your first experiments. Such a board has everything you need: an MCU, a LoRa module, an antenna, and a display. You only have to connect your sensor of choice and write a program, usually using the Arduino IDE.

You can even start without a sensor, by only transmitting a "Hello World" message to the TTN network. If your node can reach a gateway with its link budget!

> 8 parallel channels = 8 devices
>
> At 50% duty cycle = 16 devices
>
> At 1% duty cycle = 800 devices

G

> 250 bits per second / 100 = 2.5 bps
>
> Morse = 20 bps

H

LoRa STANDARDS

LoRa Alliance, which is supported by many big companies, develops the LoRa standards. This is very important for its future, as nodes manufactured by one supplier should always work with gateways made by another manufacturer.

MOVING FORWARD

LoRa fits nicely in the gap between Bluetooth for short distances and mobile internet for ubiquity. Because it uses ISM bands, it is license-free. Its unique properties are even used by satellites to offer truly global coverage (e.g., Lacuna), and by small cube satellites deployed by universities.

As LoRa's link reliability is not as high as with other technologies, it is mainly used for sensor networks where the lack of one data point is not critical. If used with sensors where transmissions occur periodically every few minutes or even hours, the power consumption can be very low, and it can run on a set of batteries for a very long time (but not for years, as its marketing has tried to suggest). To make it work, you must acknowledge the different influences on its link budget. If done correctly, you can create sizable low-cost sensor networks.

To learn more about LoRa, please visit my Youtube channel, where I delve deeper into technologies and builds for this network: youtube.com/AndreasSpiess. Good luck! ◉

TTNmapper.com shows LoRa gateways worldwide. Contribute to the map with an easy LoRa GPS tracker build like hackster.io/fablabeu/gps-mapper-for-the-things-network-ttn-lorawan-584ed7.

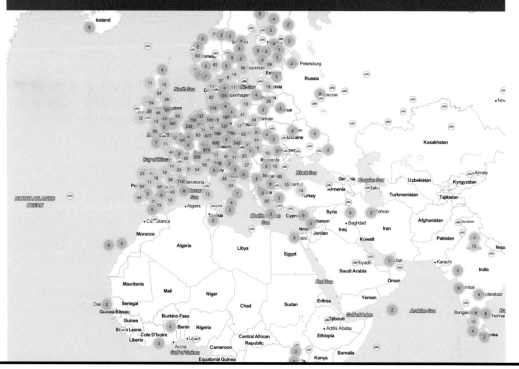

LoRa-CAPABLE BOARDS

LoRa microcontrollers have built-in radio chipsets to let you connect your DIY sensor build to a gateway without any additional hardware. Many of these providers also offer independent LoRa-capable shields, wings, and expansion boards to add to your existing microcontroller. Be sure to get the right frequency for your region!

Adafruit Feather LoRa Radio: M0 or 32u4, 900MHz or 433MHz

Arduino: MKR WAN 1300 and 1310

Grasshopper LoRa/ LoRaWAN Development Board

Penguino Feather 4260

Pycom FiPy and LoPy4

SparkFun LoRa Gateway

SparkFun Pro RF LoRa

TTGO T-Beam

PREMADE SENSOR NODES

There are many commercial LoRa sensor nodes available. If you want to assemble a system without building your own components, you'll likely find what you need already exists.

- Temperature
- Rainfall
- Flood/water
- Energy
- Pressure
- Air quality
- Open/close
- Humidity
- Gas/energy metering
- Street parking
- Mouse/rat trap
- Outdoor ultrasonic height measurement
- GPS tracking

You can find a comprehensive list at thethingsnetwork.org/ marketplace/products/devices

MORE LORA TO EXPLORE

Flex your reach with these **long-range projects**

① LORA GPS DOG TRACKER

instructables.com/LoRa-GPS-Tracker
The LoRa system is well-suited for wildlife conservation efforts, allowing for the tracking of animals through large regions. You can test this out with a GPS tracker build of your own for your favorite furry family member. **Scott Powell** steps through one approach for this with his build, using a LoRa-enabled Adafruit Feather, the Ripple network, and an Android app. The result is a compact device that attaches to his dog's collar and maps its location. An upgraded version with S.O.S. alerts can be found at hackster.io/scottpowell69/s-o-s-enabled-gps-tracker-620407.

② RAIN-SENSING LORA WEATHER STATION

how2electronics.com/lora-based-wireless-weather-station-with-arduino-esp32
This project by **Mr. Alam** in Biratnagar, Nepal builds a weather tracking system to explore the long-range, low-power aspect of LoRa. The sensor array tracks temperature, humidity, pressure, rainfall, and more, and transmits to a gateway that then pushes to a Blynk app for easy viewing on your phone.

2

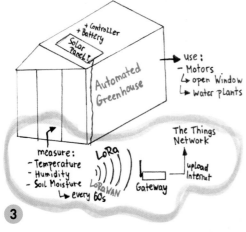

3

4

5

③ LORA-AUTOMATED GREENHOUSE

instructables.com/Automating-a-Greenhouse-With-LoRa-Part-1-Sensors-T and
instructables.com/Automating-a-Greenhouse-With-LoRa-Part-2-Motorized

A greenhouse should keep consistent heat and humidity inside. German YouTuber **GreatScottLab** designed a system to measure both of those, as well as soil moisture, and has added a motorized window-opening apparatus to help maintain proper conditions. The sensor readings and window control are transmitted over the LoRa network via his own gateway, letting him access it from almost anywhere he travels. A solar-charged battery keeps everything running.

④ LOW-POWER WATER LEVEL SENSOR/TRANSMITTER

hackster.io/Amedee/low-power-water-level-sensor-for-lorawan-the-things-network-96c877

This self-contained LoRa node uses an ultrasonic sensor to determine the water level in a not-easily accessible rainwater tank and transmit the readings. Project creator **Philippe Vanhaesendonck** from Belgium leverages LoRa's low power consumption so the sensor can run unattended for extended periods of time on battery power. The how-to instructions also note that the design can be used to measure accumulation of snow, trash, and so forth.

⑤ USING LORA WITH CIRCUIT PYTHON

learn.adafruit.com/using-lorawan-and-the-things-network-with-circuitpython

Adafruit author **Brent Rubell** has a full guide on getting started with LoRa using quick-to-deploy CircuitPython on a simple weather logging node. The guide also details how to use CircuitPython and the Adafruit-managed TinyLoRa library on Linux boards like Raspberry Pi. ●

ARMACHAT

Learn to build and flash a graphical Arduino-based device with this wireless **LoRa** "doomsday" communicator

Written and photographed by Peter Mišenko

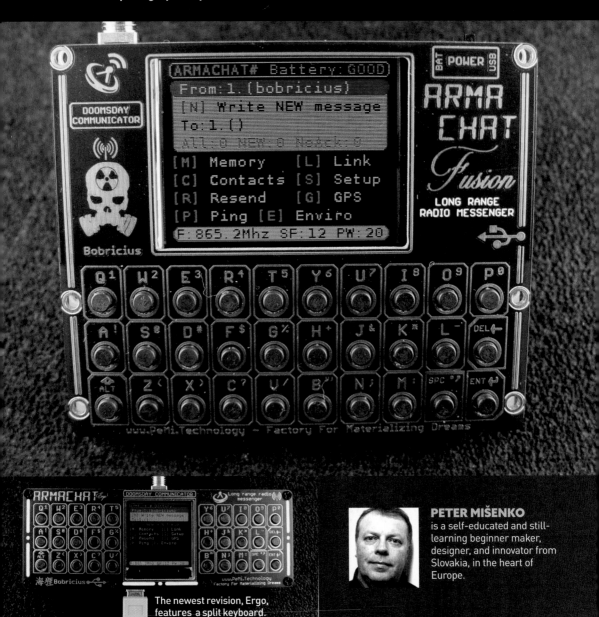

The newest revision, Ergo, features a split keyboard.

PETER MIŠENKO
is a self-educated and still-learning beginner maker, designer, and innovator from Slovakia, in the heart of Europe.

LoRa is great for long-distance sensor readings, but you can also use it for sending messages. This is my LoRa-based interpersonal communicator, called Armachat. I made it as a learning exercise to create a PCB with wireless communications and graphic displays. The device is a battery-powered, super high-tech, text-based walkie-talkie with an extremely simple and low component count (around 25 parts). I focused on using an SAMD21E18 microcontroller in an easy-to-solder TQFP32 package (same size as the popular ATmega328). Unfortunately this MCU is sold blank and needs a special Atmel-ICE device to program the bootloader — this is the hardest part of this project. Otherwise, no special manufacturing tools are required, it's all on the PCB. The front panel can also carry optional backlight LEDs for the keyboard.

I originally called it an Armageddon communicator (hence "Armachat") but I don't think it would survive any imaginable disaster.

The software is not finished yet but at this point you can compose/store and send/receive messages with delivery confirmation, and resend undelivered messages from device to device with high potential to be received.

WHY I USED THE SAMD21E18

I chose this MCU for Armachat because it is Arduino compatible, low power, has native crystal-free USB, USB OTG, a lot of flash memory for updates, MSD bootloader from Adafruit, and you can easily update firmware just by dragging and dropping like a thumb drive. Plus, it's cheap.

It's also really low power; total Armachat consumption at full power with display on is 60mA — 20mA for the LoRa module, 20mA display backlit, 20mA MCU. I'm able to boot and send messages from Armachat just using power from a Stirling engine generator. You can see this on my YouTube page, youtube. com/bobricius.

BUILD IT

The build is pretty straightforward; I've assembled 20 different prototypes just with hand soldering, and haven't had any issues.

TIME REQUIRED
1–2 Hours

DIFFICULTY
Advanced

COST
$30–$50

MATERIALS
» **Printed circuit boards (PCBs), main and front panel** tindie.com/stores/bobricius
» **LoRa radio module, RFM95W** 868MHz for Europe, 915MHz for USA, or equivalent module of correct frequency for your region
» **Microcontroller IC chip, SAMD21E18A-AU, in TQFP32 package**
» **LDO voltage regulator IC, 3.3V, AP2114H-3.3TRG1**
» **Battery charge controller IC, MCP73831**
» **Power MOSFET transistors, BSS123 (3)**
» **Tactile micro switches, DTS-63K, 1N force (30)**
» **Tactile micro switch, TACTM-35N-F**
» **Battery, Li-ion, 18650**
» **Battery holder, 18650** BHC-18650-1P
» **Micro-USB port** Molex MX-105017-0001
» **LCD display, 1.8" TFT LCD, 128x160, ST7735** such as aliexpress.com/item/32818686437.html
» **Resistors, 0603 SMD package:** 1.5kΩ (2), 4.7kΩ (7), and 10Ω (1)
» **Capacitors, 0603 package:** 100nF (2), 1µF (1), and 4.7µF (4)
» **Capacitor, tantalum, 1206 SMD package, T491A107M004AT, 4V 100µF**
» **SMA coaxial PCB edge connector** Adam Tech RF2-143-T-17-50-G or RF2-145A-T-17-50-G-HDW
» **Sound transducer, SMD** LD-BZEL-T61-0505
» **LED, 0603, red** for charging indication
» **Antenna** such as GSM-ANT-SV03 SMA Male; must correspond to your LoRa frequency
» **Power switch (side)** C&K Components JS202011AQN or Ninigi MSS-2245
» **Machine screws, M2×5mm (12)**
» **Distance bolts, M2×5mm (6)** aka threaded standoffs, male-female
» **LEDs, 0802 right angle package, green (30) (optional)** aka side view, for keyboard backlight
» **Resistor, 0603, 100Ω (optional)** for keyboard backlight

TOOLS
» **Soldering iron and solder** surface-mount-component capable
» **Screwdrivers**
» **Atmel-ICE programmer** microchip.com/DevelopmentTools/ProductDetails/ATATMEL-ICE
» **Computer with software:**
  • **Microchip Studio IDP** (formerly Atmel Studio)
  • **Arduino IDE** free from arduino.cc/downloads
  • **ArduinoCore and bootloader** github.com/mattairtech/ArduinoCore-samd
  • **Armachat firmware** github.com/bobricius/armachat; more resources at pemi.technology/ARMACHAT

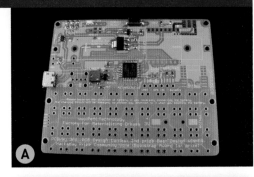

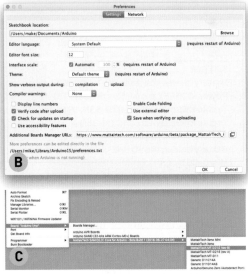

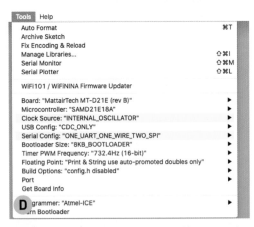

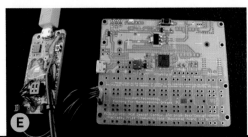

1. Assemble the PCB

Solder the CPU and all active/passive components, following the labels on the PCB, except for the display, LoRa module, buttons, and SMA antenna connector (Figure Ⓐ).

2. Program the MCU

I am using the Arduino IDE and the Atmel-ICE programmer to write the bootloader into the SAMD21E18.

In the Arduino IDE menu bar, go to Arduino → Preferences → Additional Boards Manager URLs and add the following URL: https://www.mattairtech.com/software/arduino/beta/package_MattairTech_index.json (Figure Ⓑ)

Then in the Tools menu, select Boards Manager and install MattairTech. When installation finishes, go to Tools → Boards and select the board MattairTech MT-D21E (rev B) (Figure Ⓒ).

After you select that board, you'll have some new fields to configure (Figure Ⓓ). Choose:
- Clock Source : INTERNAL_OSCILLATOR
- Serial Config : ONE_UART_ONE_WIRE_TWO_SPI

Now connect the Atmel-ICE programmer board with wires to the GND, VCC, SWC, SWD and RST pads on your Armachat board (Figure Ⓔ).

Back in the Tools menu, choose Programmer: Atmel-ICE. Then select Burn Bootloader. Now you can upload sketches to the device with USB.

Alternatively:

In Atmel Studio, write the special *ARMABOOT.bin* file, downloaded from my Github page. Then you can easily upload the firmware by dragging and dropping to your mass storage device.

3. Upload the first firmware

If you have *ARMABOOT.bin* flashed, connect Armachat to your computer via USB. Double-press the RESET button for bootloader activation.

If you have a display connected, the backlight will start flashing (breathing).

Copy *first.uf2* from my Github page to the newly displayed mass storage drive. Wait for it to finish writing, then press RESET again. If you hear a beep, the firmware is running.

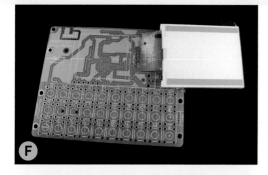

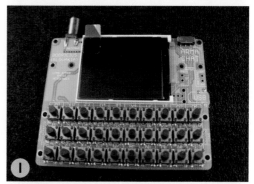

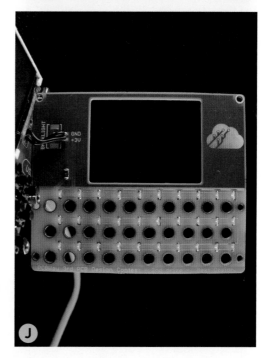

You can also create the UF2 firmware from scratch. Download the following libraries (or get them all from my Github page):

- playground.arduino.cc/Code/Keypad
- github.com/adafruit/Adafruit-GFX-Library
- github.com/adafruit/Adafruit-ST7735-Library
- github.com/sandeepmistry/arduino-LoRa
- github.com/cmaglie/FlashStorage

Plus the source code, located at github.com/bobricius/armachat. Compile and upload the code in the Arduino IDE.

4. Solder the last components

Solder the display (Figure **F**), then power up the Armachat. If you see messages, it's working. If the display is white, try pressing/holding the RESET button. If it's not working, check the flex cable connection. Once the display is working, you can remove the tape backing and stick it to the main board.

Next solder the RFM95 LoRa radio module, or the corresponding version for your region (Figure **G**). Power on the Armachat; if you see **LoRa init succeeded**, things are looking good (Figure **H**). If not, check the radio module connection.

Finally, solder the buttons and SMA antenna connector (Figure **I**).

5. Front panel backlight (optional)

This is optional. Solder in the backlit LEDs. I wanted perfect light spread so I used 30 LEDs, one for each button (Figure **J**). The front panel is then mounted with distance bolts and M2 screws.

USE IT

The communicator is controlled via shortcuts. From the main screen (Figure **K**) you can press:

- **S: Setup** To adjust communication frequency, LoRa transmission settings, and set your ID number from 0–10 (Figure **L**).
- **C: Contacts** The address book allows 10 contacts (for now) with fixed names (Figure **M**). I'm working on new software that will let you choose your name, which will automatically update to your contacts' address books.
- **N: New message** The keyboard buttons have multiple functions, switched with the Alt key. In "lowercase" mode, the Enter button sends

the message; in other modes it works as the right arrow and Del is the left arrow. The Space button creates a space, period, or comma.

After sending a message, Armachat waits for a delivery confirmation; if delivery is not confirmed, it stores the message into memory with an **undelivered** attribute.

Every unit sends an **alive** message when it powers on, and every unit that receives this message then resends any undelivered messages.

- **M: Messages** Read your new messages; and store sent/received/undelivered messages (Figure **N**).

DISTANCE TESTS

LoRa is known for its long-range capabilities. The maximum distances I've gotten with my communicator, through some small hills and houses in my village (not line-of-sight) at max power, with spread factor 12:

- 700m (about 0.43 miles) to gateway on my garden deck
- 1,500m (almost a mile) to gateway in my window, about 4m (13') above ground
- 4,300m (2.67 miles) to gateway in attic window, 10m (33') above ground

Of course this all depends on the antenna, obstacles, and radio settings.

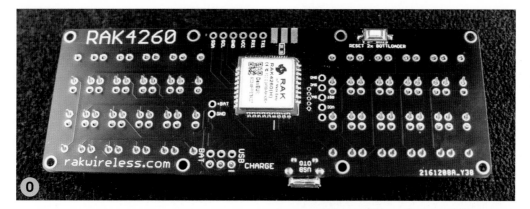

Software is now very basic but I'm intensively working on new features. There are almost no limits to routing, forwarding, mesh networking, and encryption. The device has lots of free flash memory and computing power.

CONCLUSION

Ultimately, this is a learning device to play with LoRa communication or to tinker with high-tech Arduino aspects. With it, I have furthered my skills in:

- QWERTY keyboard reading
- Creating and displaying graphics
- User interfaces
- Saving data to flash
- Sending results over radio or USB
- Working with I²C sensors
- Processing GPS data.

I'm now working on two new Ergo boards with a split keyboard that's easier to use: one with the same MCU and LoRa chips as this version, but with a 1.3" 240×240 ST7759 IPS display, and another with the RAK4260 module from rakwireless.com, which combines a Microchip ATSAMR34J18 SiP with LoRa capabilities on the chip (Figure **0**). This will really simplify the design; the board will only need the Li-ion charger, 3.3V regulator, and speaker.

And finally, I've created a compact Nano (4×4cm) version without buttons — but it does include a USB host for a PC keyboard. I'm working to set it up as a miniature repeater/receiver as a future software update. ❷

OTHER INFRASTRUCTURE-FREE COMMUNICATORS

- **Meshtastic:** A GPS- and LoRa-based mesh communicator system which you can use standalone or access via your smartphone: meshtastic.org

- **Disaster.radio:** This open source platform sets up solar-powered LoRa gateways that users can connect to via Wi-Fi to communicate off the grid. disaster.radio

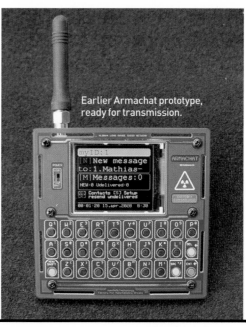

Earlier Armachat prototype, ready for transmission.

PRETTY PRINTED PATTERNS

3D print mesmerizing designs into your creations by customizing and controlling the toolpath of your first layer

Written, photographed, and illustrated by Billie Ruben

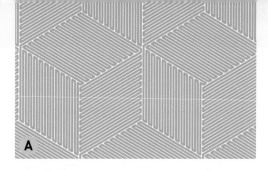

TIME REQUIRED
30 Minutes + Design Time

DIFFICULTY
Advanced

COST
Free

MATERIALS AND TOOLS
- » **FFF-based 3D printer**
- » **Filament** shiny ones like silk PLA and translucent PETG work best
- » **Vector image software** such as Inkscape (free) or Illustrator
- » **Ultimaker Cura slicer software** also free
- » **Text editor** I like Notepad++ (Windows-only) as there's a plugin that highlights G-code.
- » **3D model** to embed your pattern into the base layer
- » **My sample SVG pattern (optional)** prusaprinters.org/prints/43192-custom-patterns-for-forced-toolpath

A

BILLIE RUBEN is a maker of many kinds, she's helped run the largest 3D printing communities on Reddit and Discord, and recently started a YouTube channel. billieruben.info

B

When it comes to 3D printing, you may think you're locked into using the output of your slicer to build your creation, but there's a lot you can do with the G-code. I've started adding patterns to the bottom by controlling the direction of the lines on the first layer of the 3D print. It's done by creating an SVG image, converting it to G-code, and replacing the first layer of the model's G-code with that of the pattern.

1. MAKE AN SVG OF YOUR PATTERN

Using your favorite vector software, make your pattern as an SVG. Keep these notes in mind:

- The lines of your pattern will be lines of your print (Figures **A** and **B**)
- Pixels in your SVG equate to millimeters in Cura (Figure **C**)
- The stroke/line width in your SVG needs to match your line/nozzle width in Cura, otherwise your pattern will be over/under-extruded (Figure **D**)
- Ensure that the lines in your SVG just touch each other — no overlaps or gaps — to fill the whole area (Figure **E**)

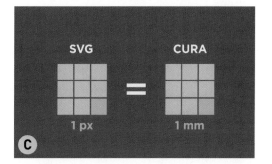

SVG   **CURA**

1 px = **1 mm**

C

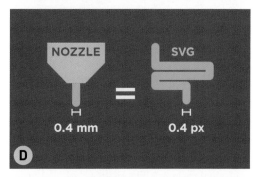

NOZZLE   **SVG**

0.4 mm = **0.4 px**

D

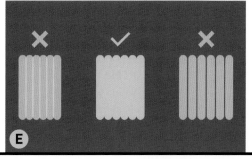

E

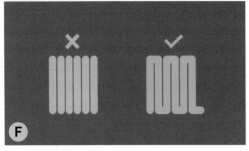

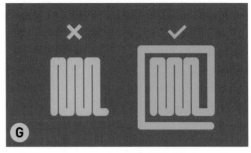

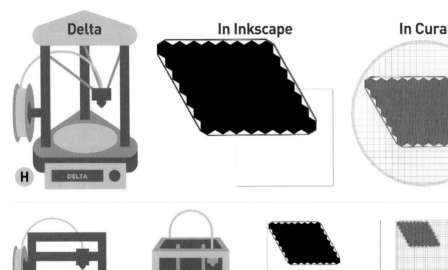

Delta · In Inkscape · In Cura

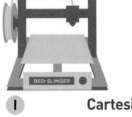

Cartesian · In Inkscape · In Cura

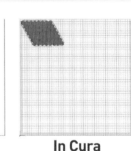

- Join all your lines up, to avoid import errors and reduce travel moves (Figure **F**)
- Add a skirt to your pattern, to prime your nozzle (Figure **G**)
- For delta printers, center your design around the top corner in your SVG as the center point of a delta is at **0,0** (Figure **H**)
- For Cartesian printers, place it inside the top left corner (Figure **I**)

2. SLICE THE MODEL TO WHICH YOU WISH TO ATTACH THE PATTERN

Open your file in Cura. Set up your normal print settings. Choose "Save to File" when finished, then close it — you won't be printing just yet.

3. INSTALL THE SVG TO G-CODE PLUGIN

Open the Cura marketplace and install SVG Toolpath Reader by Ghostkeeper (Figure **J**).

4. TURN YOUR SVG INTO G-CODE

Drag your SVG file into Cura to convert it to G-code. Check that the nozzle is starting at your skirt by hitting the Play button (Figure **K**). If it isn't, attach your skirt to the other end of your pattern (in the SVG file).

Check that the pattern covers the base of your model by overlaying two top-view screenshots in an image editor (Figure **L**), and click Save to File to save the pattern G-code.

5. REPLACE THE FIRST LAYER CODE OF THE SLICED MODEL WITH THE CODE FOR THE PATTERN.

Open your model G-code in a text editor, delete everything between the lines that say `;LAYER:0` and `;LAYER:1`, and paste in the corresponding section from your SVG pattern's G-code.

6. TIME TO PRINT

Save that spliced G-code file and print it on the printer for which your Cura is configured.

7. TIDY IT UP

After printing you'll probably have some leftover bits of your pattern sticking out. Trim these off with a pair of scissors (Figure **M**). You're all set!

GOING FURTHER

I've since made a second pattern using this method, based on a Gosper curve (Figure **N**). Next I would like to try this with a wood-grain pattern using wood-grain filament. I also think these kinds of patterns would be cool in semi-transparent objects like lamp shades.

If you get stuck anywhere, check out my video tutorial: youtu.be/zSgW0KoguXc or reach out to me on the socials: linktr.ee/BillieRuben

If you make your own I would love to see, so please tag me! Happy printing! ⦿

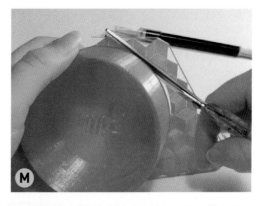

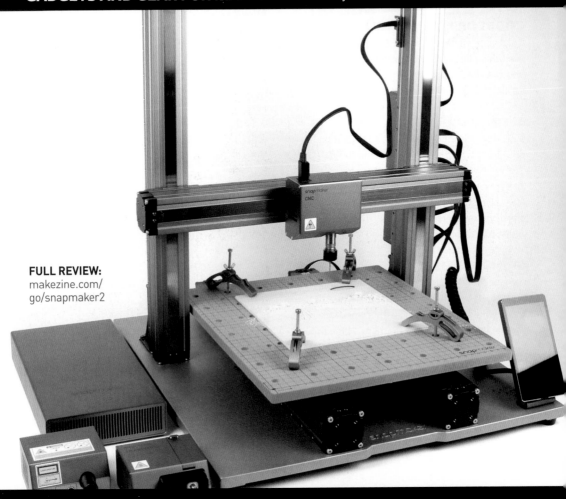

FULL REVIEW:
makezine.com/
go/snapmaker2

SNAPMAKER 2.0 3D PRINTER/LASER ENGRAVER/ CNC MILL $1,199–$1,799 snapmaker.com

Snapmaker has updated its 3-in-1 machine, and now offers two bigger options. The A150 model is still a "one-armed" design, while the new A250 and A350 (which we tested) both use a gantry due to their larger sizes.

Performance is solid. On the "fine" setting, the machine created flawless 3D-printed surfaces. The new laser module is a remarkable upgrade; instead of a weak 200mW laser, there's now a 1.6W diode, which means that engraving work can now be done in a quarter of the time. Don't expect miracles though — even thin 2mm plywood needed two passes for reliable cuts.

The laser module now also has a built-in camera, which lets you position your workpiece with millimeter precision. The machine also uses this to quickly set the laser's focus.

Snapmaker has significantly upgraded the milling head too. Instead of the clumsy drill chucks familiar from mini drilling machines, there's now a precise ER-11 collet, which is comparable in size to that of small hand-held routers. However, it only comes with one insert for ¼" tools (3.175mm shaft diameter).

All in all, the new Snapmaker 2.0 A350 is a thoroughly successful and excellently crafted device. —*Carsten Mayer*

BROTHER SCANNCUT SDX85

$349 amzn.to/35uFX60

Simply put, I love this machine. Coming from the world of vinyl cutters, the built-in ability to scan something and precisely cut around it is reserved for higher dollar machines, and often still requires software suites to set up proper registration marks. This budget hobby-level cutter does it all without the need for external software, and it does it very well.

With a 12"×12" standard cutting area, you can toss on vinyl or sticker paper and just let the machine scan it, add border spacing, and cut with fantastic precision, meaning you can produce homemade stickers that look totally professional, ridiculously easily. You can even load a roll feeder to be able to run larger quantities off of vinyl rolls. It is still a general craft cutter so you can also cut foam, fabric, etc and even change out the tool head to allow it to become a pen plotter. If I were outfitting a makerspace, this would be one of the first items I'd put on the must-have list. —*Caleb Kraft*

BEAT BARS CARDBOARD MIDI CONTROLLER PEDALS

$49 / $59 beatbars.com

MIDI's been around for ages, and is still a great tool to expand any digital music producer's capabilities. Your keyboard likely has MIDI connectivity, but you can do even more with MIDI expression pedals and footswitches. Poland-based Beat Bars makes professional-grade options for these, but their fold-your-own cardboard options should catch any maker's eye. You simply put a smartphone inside each pedal; they use the phones' sensors to transmit MIDI commands wirelessly via Bluetooth. We were surprised at how sturdy they feel and how smoothly they work. Bonus: The company has published free PDFs of the templates on their site in case you want to cut your own (or if you have to rebuild a pedal after rocking out too hard) instead of using their die-cut versions. —*Mike Senese*

BORA TOOLS NGX CLAMP EDGE SYSTEM

$199 amzn.to/3kuwOyr

Circular saw guides and track saws are the unsung heroes of lower budget woodworking. These relatively simple tools allow you to make precise and straight cuts on large pieces, and then pack away into compact spaces. There are many who rely on these tools to completely replace the table saw.

Bora Tools's NGX Deluxe set arrived on the same day my table saw died. After using this thing, I don't think I'll buy a new table saw, but rather reclaim that space. The rail clamps onto the boards, meaning that it won't slip midway through a cut, and the variations possible with the sled mean I can do angle cuts and even use it with my jig saw. I would highly recommend anyone who doesn't use their table saw daily to consider one of these. —*Caleb Kraft*

Finally!

a print magazine **+ blog**
+ newsletter
+ events

for

women

in STEM, *by*

GRL
PWR

women in

STEM.

 readers in:
50 states
25+ countries

Reinvented Inc. is a 501(c)(3) nonprofit organization that aspires to break barriers and aid the movement to get more girls involved in STEM by creating the nation's first print magazine for women in STEM.

Our mission is to reinvent the general perception of women in STEM fields while inspiring interest in STEM for young women worldwide.

Order your
Print Subscription
today!

reinventedmagazine.com

NEW BOARDS! See all our board reviews at makezine.com/comparison/boards

ADAFRUIT QT PY

$6 adafruit.com

FULL REVIEW: makezine.com/product-review/boards/adafruit-qt-py

Adafruit's new QT Py is an Arduino- and CircuitPython-capable board with a lot of features in a tiny size. It measures less than 1" square, with seven castellated pins on each of two sides at 0.1" spacing for breadboard use. Eleven of these pins can be analog inputs or digital I/O. Six can be capacitive touch inputs. One has a true analog output as a 10-bit DAC. And there are hardware SPI, I²S, I²C, and serial interfaces available on various combinations of the pins. The QT Py also includes a Stemma QT/Qwiic connector for hooking up the myriad of I²C peripherals offered by SparkFun and Adafruit. Its 32-bit chip runs roughly 4x faster and has 8x the program storage than an Arduino Uno. On the bottom are pads for an easily-solderable SPI flash chip, enabling you to add 2MB (or more!) of extra storage to the board. It also includes a reset button and Neopixel status LED. My take: It's $6 — just get it. —*Tod Kurt*

NVIDIA JETSON NANO 2GB

$59 nvidia.com

FULL REVIEW: makezine.com/product-review/boards/nvidia-jetson-nano-2gb

The Jetson Nano 2GB is a slimmed down version of its 4GB predecessor (the original Nano) with half the memory, one CSI-camera connector instead of two, and three USB connectors instead of four. For the loss of these features, users pay about $40 less than its $99 predecessor.

The Nano's secret weapon is its built-in Nvidia GPU, which makes it capable of real-time computer vision, audio processing, and other high performance applications. Even the 8GB Raspberry Pi 4 can't compete when it comes to these high-intensity jobs. Machine-learning algorithms are optimized for the GPU's parallel processing power. If your next embedded project involves AI or a lot of image processing this may be the board for you.
—*Kelly Egan*

OVER THE TOP

DOUBLE DECKER

Belgian construction-innovation firm **Kamp C** raised the roof on a 26-foot-tall, 969-square-foot concept dwelling built entirely in one piece using the world's largest 3D concrete printer. It features two stories and concrete overhangs — firsts in printed buildings. The firm is now assessing long-term viability. More: kampc.be/c3po_eng

KAMP C and Jasmien Smets